In Search of Us

TWELVE ADVENTURES
IN ANTHROPOLOGY

———•———

Lucy Moore

Atlantic Books
London

First published in hardback in Great Britain in 2022 by Atlantic Books,
an imprint of Atlantic Books Ltd.

This paperback edition first published in Great Britain in 2023 by
Atlantic Books.

Copyright © Lucy Moore, 2022

The moral right of Lucy Moore to be identified as the author of this work
has been asserted by her in accordance with the Copyright, Designs and
Patents Act of 1988.

10 9 8 7 6 5 4 3 2 1

A CIP catalogue record for this book is available from the British
Library.

Paperback ISBN: 978-1-78649-917-2
E-book ISBN: 978-1-78649-916-5

Printed in Great Britain by Clays Ltd, Elcograf S.p.A.

Atlantic Books
An imprint of Atlantic Books Ltd
Ormond House
26–27 Boswell Street
London
WC1N 3JZ
www.atlantic-books.co.uk

In Search of Us

Lucy Moore is an author and broadcaster whose nine books include the bestselling *Maharanis: The Lives and Times of Three Generations of Indian Princesses*. She has written for the *Sunday Times*, *Observer*, *Vogue* and *Harper's Bazaar*, and has presented series for the BBC and Sky. For further information, please visit the author's website: www.lucymoorebooks.com

'In this skilful summary of the early years of anthropology between 1880 and 1939, Lucy Moore reveals a veritable tangle of turf wars, power scrambles and sexual bad behaviour... Moore's fluent account confirms that there is always room for a new view, especially when it is as well done as this one.'
Sunday Times

'Moore doesn't sugar-coat her protagonists' many prejudices, their cavalier treatment of their indigenous subjects, or the problematic history of their discipline. But though she summarises their scholarly views, the main pleasure of her book lies in its celebration of a dozen colourful, unconventional, free-thinking lives.' *Guardian*

'The story of anthropology's early pioneers lies at the heart of this joyfully narrated history of a scientific field that, at its best, opens our minds to the rich kaleidoscope of human experience... [A] gripping collection of life stories.'
Literary Review

'Entertaining... Told with a novelistic eye for the character-revealing anecdote.' *Spectator*

By the Same Author

Lady Fanshawe's Receipt Book:
An Englishwoman's Life During the Civil War

Nijinsky: A Life

Anything Goes: A Biography of the
Roaring Twenties

Liberty: The Lives and Times of Six Women
in Revolutionary France

Maharanis: The Lives and Times of
Three Generations of Indian Princesses

Amphibious Thing: The Life of a Georgian Rake

The Thieves' Opera: The Remarkable Lives
and Deaths of Jonathan Wild, Thief-Taker, and
Jack Sheppard, House-Breaker

Con Men and Cutpurses: Scenes from the
Hogarthian Underworld

'One is constantly wondering what sort of lives other people lead, and how they take things. I am quite obliged to Mrs Cadwallader for coming and calling me out of the library.'

Dorothea Brooke in George Eliot, *Middlemarch* (1872)

<hr />

It thus suffices for history to take its distance from us in time, or for us to take our distance from it in thought, for it to cease being internalizable and to lose its intelligibility, an illusion attached to a provisional interiority. But that does not mean that I am saying that man can or should free himself from this interiority. It is not in his power to do so, and for him wisdom consists in watching himself live it, knowing all the while (but in another register) that what he is living so completely and intensely is a myth, which will appear as such to men of a future century, which will appear as such to himself, perhaps, a few years hence, and which to men of a future millennium will not appear at all.

Claude Lévi-Strauss, *Pensées Sauvages* (1962)

Contents

Introduction

——•——

This book follows twelve European and American anthropologists over a period of fifty years, from the 1880s to the 1930s, as they lived with and systematically observed indigenous people they called savages, in what were then considered the most exotic corners of the globe: the wind-swept snowfields of the Arctic, the impenetrable jungles of Brazil, the sawmills and illicit roadside bars of the American Deep South. Each chapter looks at an anthropologist (or two) during a specific moment in the field, at a formative moment in their career, examining the lessons they learned and the way they communicated them on their return. Taken together, they create a broader narrative about the story anthropology as a whole was telling contemporary Western people about themselves – a story that fundamentally shaped the way we as individuals and societies look at one another today. Over this first half of the tumultuous twentieth century, anthropology, by endeavouring to explain human beings and their cultures, offered the possibility – at once thrillingly contemporary and strangely comforting – that hidden in the way people interacted were universal truths that could be applied to their own rapidly changing world.

Interest in the 'primitive', the Other, was blossoming in the developed world during this period, feeding into and stimulated by the emergence of anthropology as an intellectual, cultural and political movement. This was reflected in the broader trends of the day: the serene Tahitian paintings of Paul Gauguin; an early-twentieth-century soundtrack of African American blues, jazz, ragtime and spirituals echoing out from modern Victrolas; the African-inspired masks of Pablo Picasso; and war-weary Bloomsbury bohemian Gerald Brenan escaping shell shock by living with and writing about the people of remote Andalusia. While Carl Jung was researching his theory of archetypes, he visited the pueblos of the south-western United States, which also sparked the imagination of Aldous Huxley; the Savage Reservation Lenina and Bernard visit by Blue Pacific Rocket in *Brave New World* (1932) is one of the more memorable and unsettling twentieth-century visions of the future.

This period in the history of anthropology as a discipline was marked by its new devotion to fieldwork, which can be defined as empirical research performed on the ground, in the field, rather than in a laboratory or library. Originally used for research in the natural sciences, over the late nineteenth and early twentieth centuries it was enthusiastically adopted by the new social sciences and, particularly, anthropology, the study of humankind, with work in far-flung locales becoming the indispensable initiation into the discipline and an essential part of its mystique. As Alfred Radcliffe-Brown would put it in 1922, after his time with the Andaman Islanders and in Australia, 'It is only by actually living with and working amongst a primitive people that the social anthropologist can acquire his real training.'

Before fieldwork permeated the discipline, it was divided into data-collectors, often called ethnographers,* and theorisers. 'Observation and comparison to be kept strictly apart and carried out simultaneously by different classes of workers,' wrote Sir James Frazer in his notes for an introductory lecture to young anthropologists. Formalising fieldwork allowed the same person (increasingly regularly called an anthropologist) to observe behaviour and later to describe and analyse it. As Alfred Haddon, one of the original British fieldworkers from the generation below Frazer, would explain, 'the most valuable generalisations are made ... when the observer is at the same time a generaliser'. Anthropology was also largely a literary and philosophical exercise until the influence of the sciences, and field research in particular, gave it a more practical focus.

Eager to learn more about their own society and motivated by the desire to improve it, these young social scientists turned to apparently simpler, 'primitive' societies with the idea that they

---

* Dictionary definitions: anthropology, the study of human societies and cultures and their development; ethnography, the scientific description of peoples and cultures, their customs, habits and mutual differences; ethnology, the study of the characteristics of different peoples and the differences and relationships between them.

provided 'laboratory conditions' (Margaret Mead's phrase) for studying human culture untainted by modern civilisation: a bit like returning to Eden to study Adam and Eve before they had bitten into the apple or, to be more critical, like studying animals in a contained enclosure. It would never have occurred to them that their behaviour might be as strange as that of the 'savages' they were observing. At the time, there would have been no question: the pith-helmeted anthropologists with their intrusive questions were advanced and benevolent and their subjects, the indigenous people they 'studied', bare-skinned and adorned with feathers and shells, undeniably savage. Today that distinction is very much less clear; indeed, these labels no longer apply, and the work of these pioneering thinkers, whatever we may think of how they did it, is the primary reason for that change.

At some point over these decades, anthropologists began to realise that fieldwork posed more questions than it was able to answer. Just as the astronaut's journey into space means nothing unless she can come back to earth and tell us what she has seen, so the anthropologist cannot simply observe other people but must interpret their society and communicate her vision. As the saying goes, she must make the strange familiar and the familiar strange, all the better to understand both. Anthropology, as Robert Lowie would observe, should mean more than the study of 'savages': at its best, it ought to be 'an insight into human culture in all its reaches'. The ground-breaking social scientists in this book had intended to explain the primitive world to the civilised one, but they ended up redefining what it meant to be both civilised and savage.

Anthropology in general, and fieldwork in particular, acts like double-sided mirror glass, if such a thing is possible. The anthropologist sees her own society reflected darkly back at her when she looks at another society and the society she observes begins to see itself through her eyes. When these societies are non-literate – or, as in the case of the great Middle and South American cultures when they first encountered Westerners in the sixteenth century, had had their written histories destroyed – very little survives to tell us how they felt about being 'observed' or what contact with this strange new culture meant to them. If only one could read this book in reverse, written from the point of view of the people being studied, rather than the studiers; instead their responses must be inferred from the devastating outcomes of their contact with the developed world: the ravages of disease, alcoholism, declining birth rates, fatal lassitude – what one of the anthropologists below would define as 'racial suicide'.

But the lack of written history among the cultures they studied provided anthropologists with a key justification for their work. Using 'scientific' methods, they sought to create a record of vanishing cultures that, even at the start of this period, they recognised contact with their own culture with all its brittle sophistication would inevitably destroy. 'The great literate civilisations of the world are able to bequeath to subsequent civilisations their art and literature, their laws and inventions, but primitive people without a written language, who fashion their tools and weapons from wood, and build houses which combine perfect adjustment to sun and hurricane with a life expectancy of ten years, have no way of making such a contribution to history,' wrote Margaret Mead of her desire to commemorate 'people whose grace lies in the way they *sing* their songs, and not in the songs themselves'. One of the things that marked Mead's generation of anthropologists was their determination to understand and communicate these cultures to their own society, for its own benefit, and only secondly for that of the cultures they observed.

People have been conducting anthropological studies through fieldwork in their various ways since antiquity, more or less analytically observing other people and seeking to learn from their different habits and ways of life. Herodotus is often called the father of history but his *Histories* (dating to the fifth century BC) include numerous ethnographic observations; like the slightly younger Thucydides, another claimant to the father of history title, he evidently did his own field research. Although Pliny the Elder described his own *Natural History* (*c.* AD 77) as taking all of the natural world as its subject, including specifically anthropological chapters on humans and their history, art, medicine and magic, agriculture, metallurgy and mining, he was more armchair compiler and theoretician than fieldworker.

In later centuries, Michel de Montaigne in France and Francis

Bacon and Thomas Browne in England (among many others) demonstrated notable curiosity about the habits of humanity, as well as the broader natural sciences, but not until the eighteenth century did scholars turn with application and focus to the study of humankind. Anthropology, as it would become known, was the perfect discipline for the Age of Enlightenment, with its dual focus on scientific methods and the rights and workings of the individual within society, although early ethnographers tended to come to a study of man from a broad range of other subjects, from philosophy and politics through physiognomy and palaeontology. Carl Linnaeus classified and named the species of the known world, including humans, while the Comte de Buffon sought to examine and describe them; Jean-Jacques Rousseau and Immanuel Kant delved into humanity's heart and soul.

In 1719, the German physician, naturalist and geographer Daniel Gottlieb Messerschmidt set out from Moscow tasked by Peter the Great with exploring Siberia. He spent eight years surveying the vast and uncharted area, recording his observations on the people he met there as well as the area's flora and fauna in notebooks, maps and drawings, and collecting rare and exotic items, including the first known woolly mammoth fossils. Messerschmidt never published his findings and died in poverty but he was followed by Johann Georg Gmelin and Georg Wilhelm Steller on extensive Russian state-funded scientific expeditions all the way to the Bering Straits, which involved many hundreds of people over the next decades.

Captain James Cook set out to explore the Southern Seas in the 1760s and 1770s in a similar vein of state-sponsored discovery. Joseph Banks, the young polymath who travelled with Cook on that first voyage on the *Endeavour*, left a celebrated description of their contact with the inhabitants of Tahiti during a three-month stay there. Although he noted the generosity with which they were welcomed by the islanders, his account of their arrival betrays a sense of privilege that jars today. As they approached land, canoes came out to meet Banks, Cook and their companions and escort them ashore, where, in shaded groves, they exchanged green boughs of friendship with the Tahitians: 'in short the scene we saw was the truest picture of an arcadia of which we were going to be kings that the imagination can form'.

Banks's journal reveals various gaffes, like almost immediately insulting the chief's wife ('ugly enough in conscience') by ignoring her polite welcome to flirt with 'a very pretty girl with a fire in her eyes that I had not seen before', but as an observer, he was open-minded and kind, as well as being inquisitive (some might say nosy), having a gift for languages and lacking any sentimental preconceived ideas about the 'noble savage'. He noted the exquisite tattooing of his new friends, their navigational skills, lovemaking (directly and with enthusiasm), mourning customs, craft, food and hygiene.

Religion and ritual were more opaque. Curious to see that Tahitian women never ate with men, to the point of throwing away food if a man had inadvertently touched the basket containing it, he asked them why but 'they gave me no other answer but that they did it because it was right, and expressed much disgust when I told them that in England men and women eat together'. What was even more interesting was that when they were alone with Banks and his companions, they were willing to eat with them – as long as the Englishmen promised not to tell.

The world's first anthropological society, the short-lived Société des observateurs de l'homme, was founded in Paris in 1799 by the poet and educator Louis-François Jauffret with the motto '*Connais-toi toi-même*' ('Know yourself'). It counted the zoologist and palaeontologist Frédéric Cuvier, the physician and psychiatrist Philippe Pinel, and the explorer Antoine de Bougainville among its members. The philosopher Joseph-Marie Degérando was another associate who systematically observed 'savage people' for the clues they offered to human nature. Another member, the young explorer François Péron, working in Australia, sought to investigate Rousseau's theory that indigenous people might be stronger and healthier than Europeans, their physical health in inverse proportion to their 'moral development'.

Ethnological societies were founded over the next decades in Paris (1839), New York (1842) and London (1844), and anthropological ones in Berlin (1869) and Vienna (1870). Advances in scientific knowledge in related fields – notably the publication of *On the Origin of Species* in 1859 but also the development of archaeological excavation and the better understanding of ancient civilisations, for example with the deciphering of the Rosetta

Stone – further stimulated anthropological research and the development of the discipline. In 1858, prehistoric human bones were discovered in Brixham Cave on the coast of Devon alongside the remains of extinct species including aurochs and woolly mammoth. It was becoming increasingly clear that humanity had developed over a very long time and that long-vanished or vanishing cultures merited study and reassessment, although the assumption was that these early humans had been markedly different from modern people.

The dominance of the concept of evolution in nineteenth-century scientific thought coloured ethnology, too. It was assumed 'that all cultural systems ... progress slowly and unalterably through the same invariable stages of development', with non-literate societies at the bottom and sophisticated, industrialised civilisations at the top. All primitive people, it was believed, worshipped ancestor spirits and totems; there were no families, but women and goods were held in common by the men of the group; marriage, as it developed, was a form of exchange between groups. 'Savages' could be seen as little more than animals. Gradually as a society became more sophisticated it would move through various predetermined stages before ending in a version of top-hatted, nineteenth-century European society: a robust, monotheistic, monogamous, patriarchal, hierarchical world in which everyone knew his place and status.

Among the first academic anthropologists in England was Sir Edward Burnett Tylor, Oxford's inaugural Reader in

Anthropology (in 1883) and the author of the influential *Primitive Culture* (1871), who formulated a theory of the evolution of society from savagery through barbarism to civilisation. His readership was funded by General Augustus Henry Lane-Fox Pitt Rivers, who also bequeathed to Oxford University the extraordinary archaeological and ethnological collections that would become the Pitt Rivers Museum. Sir James Frazer spent his career largely at Cambridge, writing twelve magisterial volumes of *The Golden Bough* (published between 1890 and 1915), a comparative study of world mythologies in which he outlined the development of religious thought from animism or magic through organised religion to science and heavily implied that Christianity was a sacrificial cult like any other.

As a young man interested in 'primitive' and prehistoric humans, Tylor had encountered tribal people in Mexico, but Frazer seldom vacated his armchair. When William James asked him if he had ever met any 'natives', Frazer is said to have exclaimed, 'Good God, no!' Instead, he and Tylor made use of the accounts of colonial administrators, explorers, traders and missionaries, amateur ethnographers across the British Empire, to gather information and items of material culture they could analyse in their ivory towers. The French had a word, *coutumier*, for the official descriptions and inventories of indigenous customs in the areas under their colonial control in Asia and Africa, sent back to their university ethnographic departments. Without substantial overseas territories, anthropologists in the United States focused on the 'primitive' people *in situ*: first Native Americans and then, with a slightly different twist, African Americans.

Social evolutionism appeared, to the white European adherents of the theory, to justify 'the presumed superiority of white-skinned civilised men to dark-skinned savages by placing them both on a single developmental ladder extending upwards from the apes'. Western man's suspicion that all primitive societies could be judged against his own and found inferior seemed to be confirmed by social Darwinism. Eugenicists and racists flourished among nineteenth-century anthropologists, who allocated formulaic characteristics to the so-called races of man (Asian or 'yellow' people were thought to have genius or cunning, black people possessed soul but were childlike, white northern Europeans were energetic and honest) and justified structural inequality – slavery and genocide, at its worst – with the idea that black people and Jews, for example, were biologically inferior to whites, even subhuman. Unchallenged Nazi anthropologists would run away with these ideas in the 1930s in the darkest possible ways.

It was morally wrong, concluded some, to sustain the weakest in society, who should rather be allowed to wither away, leaving the strongest and fittest to prevail. Not everyone thought this way – some early anthropologists argued against calls for racial 'purity', claiming that 'mongrels' or miscegenation, the mixing of races, would strengthen human stock – but a confusion about biological and social processes permeated the early decades of the discipline. Hair texture, skin colour or nose size apart, a large part of the work of the anthropologists in this book involved demonstrating that differences between peoples were cultural rather than biological.

It was onto this stage that the first modern fieldworker stepped, snugly dressed in a seal-fur parka (the Inuit word for coat). Franz Boas, born in 1858 and brought up and educated in Germany, had been deeply influenced by his mentor, Adolf Bastian, an early proponent of fieldwork and in 1873 one of the founders, and first director, of the Ethnological Museum of Berlin, who believed that all human beings possessed the same mental capacities, and by Rudolf Virchow, a gifted physician and polymath, who passionately rejected scientific racism. (Unable to countenance the idea that man was descended from apes, Virchow also challenged the theory of human evolution.) Boas's 1883 expedition to the Canadian Arctic is the first episode of modern anthropological fieldwork and the starting point for the academic discipline of anthropology as it is practised today. Over the next fifty years, as Boas and his successors set forth into the field, in the parts of the world least known by developed nations, they sought to understand and engage with people they saw, initially at least, as savages. It is the change in the way they viewed the people they met with which this book is largely concerned.

Boas would become the first champion of the anthropological concept of cultural relativism, which holds that although people see the world as they are conditioned to see it, judging it through culturally acquired norms, no society is intrinsically better or worse, higher or lower, more or less civilised than any other. Beneath a top hat or a penis sheath, one human is the same as another human. Another anthropological insight, linked to the psychological work of Carl Jung and Sigmund Freud, was an awareness of the survival of 'primitive' urges in modern life, or rather the understanding that the primitive or the unconscious was an ineradicable part of human nature. Finally, in the academic work done by these anthropologists, the sense that human existence was progressing towards an ideal, predetermined end was undermined by a replacement of dynamic analysis – in which societies were compared with one another as they evolved over time – with synchronic or static analysis, studying a snapshot of a society: life as it functions. Looking at all societies as if they were suspended in time eroded the idea that one – European civilisation – was predominant. There was no ideal endpoint for humanity to reach, or even to aim for.

If, as Eric Hobsbawm has suggested, the twentieth century was marked by a revolution in social relations, these driven young anthropologists and their work were at the heart of that change. A 'sudden, moral and cultural revolution, a dramatic transformation of the conventions of social and personal behaviour': new virtues – pluralism, scepticism, tolerance and responsiveness – were replacing deference, faith in organised religion and social hierarchy, and acceptance of the way things were or had to be. Pioneers of a new way of thinking, these fieldworkers redefined ideas of race, gender, sexual orientation, parenting, class and religion. The legacy they would bequeath future generations was teaching people to try and look at one another, as societies and individuals, with eyes washed free from prejudice.

This survey – to use an anthropological term – is intended to be biographical rather than anthropological. As the great historian of the discipline George Stocking observed in 1968, despite their keen interest in 'informal oral histories' – otherwise known as gossip – anthropologists have tended to view their discipline through preoccupations with 'general developmental framework', 'theoretical or methodological relevance' and 'interpretative centres of gravity', but I'm concerned with life stories, not academic critique. In these sketches of my subjects' life and work, I want to address the central question of intellectual history as Stocking posed it: 'What was bugging them?'

Emmanuelle Loyer speculates in her magisterial biography of Claude Lévi-Strauss that anthropology provided him with a 'biographical accident' that 'was one possible way to reconcile life and writing, scholarly work and adventure, the sensory and rational worlds'. This was true of each of these anthropologists in their different ways. And yet, as Lévi-Strauss observed of his own conversion from philosophy to anthropology, it 'is one of the few genuine vocations. One can discover it in oneself, even though one may have been taught nothing about it.' He liked to compare the anthropologist returning from the field with an adolescent returning from an initiation ritual, having sought and found wisdom that will guide her through adulthood and contribute to the well-being of her tribe. He described the young fieldworker living outside her social group for a predetermined period, exposing herself to unfamiliar and often extreme conditions, and returning to share her insights and be garlanded by her community and welcomed into maturity. This was an experience all these anthropologists would have recognised.

They can be seen in their own ways as outsiders, who viewed their own societies with a wary gaze. They might be immigrants, recent arrivals to a new country; slightingly treated as inferior, despite ostensible equality, like the several Jewish and one black scholar; women struggling to achieve professional parity and personal independence; some, men and women, with (or perhaps wanting to have) private lives out of tune with their contemporaries'. The lessons they learned about freedom from ethnocentrism and cultural relativity came naturally to them: they were halfway there before they ever stepped into the field. Innovative, unconventional new disciplines, of which anthropology was the central one, offered academics like them 'paths to success off the beaten track', an opportunity to shine in their own way on the edges of the establishment.

This outsider status gave each of them an inner motivational force, a fiery sense of social justice. Throughout his life Franz Boas was trying to overcome the sense of inadequacy internalised by growing up Jewish in an anti-Semitic Europe; Edvard Westermarck kept his private life so private that the best hints at it are only his bachelor status and the closeness of his lifelong friendship with Sîdi Abdsslam; Zora Neale Hurston had to strive and scrap and occasionally lie for every chance she was given. It made them tenacious, persuasive, effective communicators, writers and, in some cases, campaigners. 'Never doubt that a small group of thoughtful, committed citizens can change the world,' famously observed Margaret Mead. 'Indeed, it's the only thing that ever has.'

Having said that, they were only human and their ambitions were never entirely altruistic. Anthropologists who had spent a year or so with indigenous people might come home with enough material to formulate an intellectual credo and define their career. While they demanded freedom from prejudice at home, in the field they studied people who could not escape their gaze. Almost all lived in the field as a matter of course with a small army of servants, kitchen boys and bearers they barely mentioned and some revelled unashamedly in being seen as special, almost exalted, by their 'native' hosts. When they left the field, having expended all their energy and charm on procuring the information they wanted, they seldom looked back. 'Where are you now?' wrote Mead's Samoan informant and friend, Fa'amotu, two years after her departure. 'We haven't received a single letter from you. Why haven't you written to us? I wish you would write to us. We love you so much and we still remember you.'

Unconscious bias was not a phrase with which they would have been familiar but it is a concept that can be applied to all of them; it's interesting to consider the concepts they didn't think to question, especially when they imagined they were in the business of questioning everything. But it's also too easy to look back and judge their words – and lives – by contemporary standards, forgetting how ground-breaking they were, often how courageous, and how influential in creating the kind of world in which their own words could be used against them. One of the lessons anthropology teaches is that individuals must be seen in the context of their own society; it would be unfair to condemn these fieldworkers for falling below the standards of behaviour accepted today.

'What did the chief say to the anthropologist after a long afternoon of conversation?' runs one of the rusty jokes about the discipline. 'But enough about you, let's talk about me.' Not until the end of the period covered by this book did anthropologists begin to grapple with the idea that far from being neutral observers who studied people almost invisibly, in fact they had a profound impact on the groups generous enough to permit them access. One of the notable things about this period is that these anthropologists largely lacked self-awareness, our pervasive postmodern sense of ironic detachment – stemming from that very concept of cultural relativism they espoused.

Over the decades since the Second World War, anthropology has had to contend with deeply troubling accusations: at the least that, unwittingly or not, in its interdependence with colonialism it exploited the people it liked to believe it was helping. Even the foundational concepts of 'discovery', 'exploration' and 'knowledge' are today critiqued as problematic. The language alone raises questions. More recently, indigenous people being 'studied' and 'observed' have been described as informants, hosts or collaborators; some of the anthropologists in this book called them friends. Casually racist terms sit in anthropological narratives alongside powerful arguments for greater understanding and generosity of spirit. Perhaps a Haitian saying best sums it up: 'When the anthropologist arrives, the gods depart.'

Fieldwork may have been for many of these academics an excuse for adventure but, by the end of their periods in the field, quite a few had antagonised their hosts and were longing for home. At some point, write Kathleen and Billie DeWalt, authors of a recent guide to fieldwork, 'we just want to be able to defecate in private, throw the toilet paper in the toilet, look another person in the eye, [and] communicate effectively without being laughed at by the people with whom we are trying to communicate'. Others embraced immersion, returning to the field again and again throughout their careers, following Margaret Mead's counsel that 'the way to do fieldwork is never to come up for air until it is all over'. Alfred Radcliffe-Brown offered characteristically sibylline advice, telling a student at the University of Chicago in the 1930s to 'get a large notebook and start in the middle because you never know which way things will develop'. Like the best anthropological axioms, it can be applied to almost anything.

CHAPTER ONE

The Pioneer

Franz Boas on Baffin Island, 1883

———•———

'It is funny how everybody thinks I am making this trip for fame and glory,' wrote the young geographer Franz Boas to his fiancée, Marie Krackowizer, in July 1883. It was a letter he hoped she might receive one day; he had no expectation of seeing her again for at least a year, possibly ever. Unable to land, his ship was hovering in the icy seas beyond fog-shrouded Baffin Island in the Canadian Arctic. 'You know that I strive for a higher thing and that this trip is only a means to that goal ... Empty glory means nothing to me.'

That April, Boas had left Marie behind in Stuttgart without declaring himself to her. They had met initially on a walking holiday in the picturesque Harz Mountains two years earlier and Boas, conscious that he was about to set off for the edge of the world, had feigned an appointment in Stuttgart to see her again. They had only a few weeks together and he thought it unfair to ask her to become engaged to a man about to embark on a long and perilous journey. But, unable to depart without confessing his feelings, he had written to her. After three weeks' passionate correspondence, she had accepted his proposal. Now he travelled with the dreamy strains of the song he had played for her in Stuttgart when they fell in love, one of Robert Schumann's four melancholic *Nachtstücke*, echoing in his head, and his loneliness was eased knowing she was waiting for him.

The expedition he hoped would make his name was a year-long geographical exploration of the area around Cumberland Sound in the north-east of Canada; the United Kingdom had just transferred its claims of sovereignty over the Arctic region to the newly established Dominion of Canada. Christian missionaries, harbingers of Western 'civilisation', wouldn't arrive until ten years later but the area's indigenous inhabitants were already in contact with European whalers and dying from the syphilis they had brought with them. Accompanied by Wilhelm Weike, a family servant paid by his father, Boas was considered (and considered himself) 'alone'. This type of detailed study of a limited region over such a long period of time, by one man living largely off the land, was entirely new.

In July, three months after bidding Marie farewell, he could see his destination but the frozen sea was almost impassable and he was struggling in vain against seasickness. At one point it took four weeks to travel fifteen nautical miles. 'All we can see, looking landwards, is a desert of ice, shoal after shoal, field upon field, broken only by an occasional iceberg.' It must have been hard to sight glory, empty or otherwise, in the white distance, but Boas was undaunted.

At last they landed in Kekerten, a small island just off Baffin, having rocked offshore for two months, and Boas settled into the rough Scottish whalers' station that would be his base for the next year, making his first contact with the Eskimo,* as he called the Inuit people. He'd brought gun cartridges, needles, tobacco and molasses, which he knew they valued highly, to trade for information and guiding. He had also brought basic medical supplies (opium, quinine, ammonia and turpentine) and he was delighted when his new companions began to call him Doctora'dluk, or Big Doctor. During the day, he focused on charting and cartography ('almost the whole of Kekerten is drawing maps for me'); at night, his new friends sang, told him stories and taught him gambling and games. He noted that the women especially were skilled at string games or cat's cradle, making 'figures out of a loop'.

But Boas found it hard to accustom himself to some aspects of his new life. He was repulsed by the strong smell of the caribou-hide tents and the taste of the Inuits' staple foods, seal meat and gulls' eggs. His attitude was patronising; he wrote, in his first months in Kekerten, of looking after his Inuit hosts as if he was the adult and they his children, forcing himself to eat

---

* Eskimo was the word in then common use for all indigenous polar groups; in fact it refers to two related peoples, the Inuit and the Yupik, of Alaska, Greenland, Canada and Siberia. Although the word Eskimo is still in use, some Inuit people consider it offensive because of its disputed etymology.

the same food as them 'so that I could always say, I have it no better than you'.

Marie was constantly in his thoughts. 'Opposite me rose the steep and threatening black cliffs, the rapids we had crossed that afternoon rushed and roared at my side, and in the far distance shone the snow-covered mountains. But I saw only you, my Marie. You and the noble beauty of my surroundings made me conscious of the immensity of our separation.' The immensity of their separation, the wild romance of his uncharted environment, the rigours of his work and his soaring ambition: Boas's letter-diaries to his beloved reveal the relish with which he faced the challenges ahead of him. At twenty-five years old, he saw this expedition, which he had planned for a year and dreamed of since childhood, as his chance to prove himself to the world.

Franz Boas was born in Minden, Westphalia, in 1858, one of six children but the only boy in his family to survive to adulthood. His parents were prosperous merchants, selling, importing and exporting high-end millinery, part of a small, long-standing Jewish population in the ancient town. Secular, intellectual and idealistic, they had supported and been disappointed by the revolutions of 1848, after which Sophie Boas's brother-in-law, Abraham Jacobi, had been exiled. Uncle Jacobi, a friend of Karl Marx, became a doctor in New York, specialising in public health and paediatrics. His nephew, the young Franz, was gifted, particularly in mathematics, but he gave up his hopes of becoming a professional pianist only reluctantly to focus on his academic work.

Laurels shimmered ahead of him, tantalisingly out of reach. For a young Jewish boy in nineteenth-century Prussia, no matter how talented, a public career of any kind was an impossible dream unless he openly renounced his faith and converted to Christianity. Aged sixteen, he wrote to his sister, who had chided him for his ambition, 'It seems terrible to me to have to spend my life unknown and unnoticed by people. But I am afraid that none of these expectations will ever be fulfilled. I am scared myself of such thirst for glory, but I cannot help it.'

In the late 1870s and early 1880s, Boas moved between three universities – starting at Heidelberg and ending in Kiel, via Bonn – where he studied mainly physics, philosophy and geography but also attended courses in mathematics, chemistry, botany, comparative anatomy, astronomy, folklore, geology and biology. He wrote his dissertation on the perception of the colour of water, an early expression of his lifelong intellectual 'desire to understand the relation between the objective and subjective worlds'. On leaving Kiel (the least well-regarded academically of the three universities he attended but where one of his sisters was living), he received the second highest grade of his year.

At least five times he met fellow students in duels. The first incident was almost frivolous, a dispute over an overplayed rental piano, but later he challenged anti-Semitic students seeking to exclude Jews from university life – 'the damned Jew baiters', as he called them in a letter home after one bout, warning his parents to expect him scarred by 'a few cuts, one even on the nose!' Duelling, or *Mensur*, was a craze in German universities at the time, with established conventions including padded clothes and goggles for both insulter and insulted, seconds, an umpire, and a surgeon on hand, and Boas was far from the only student proudly to boast facial scars, or *Schmiss*, from an opponent's épée, but his impatience to defend his honour reveals the depth of his internalised feelings of inferiority as a second-class citizen. Later, in his declaration to Marie of his desire to 'live and die for … equal rights to all, equal possibilities to learn and work for rich and poor alike', it is not hard to attribute the roots of his egalitarianism to this ingrained and unwelcome sense of inadequacy; its scars, at any rate, were there for all to see.

The empty Arctic might seem an unlikely place to prove oneself, but Boas's childhood had coincided with a period of German polar exploration and, with Alexander von Humboldt a childhood hero, he had always yearned to travel. The geographical expedition he planned appealed to him partly because it was interdisciplinary, requiring talents across a variety of subjects, but also because it was for him alone: his challenges to be faced and his glory to be achieved. He spent a year applying for grants and writing articles to raise funding and teaching himself everything from meteorology and the new technology of photography to Inuit language and linguistics, to ready himself for the journey that would decide his fate.

When winter came to the Cumberland Sound, Boas travelled with the Inuit, researching their hunting and migration routes.

That year the snow was unusually soft and deep (it is to Boas that we owe the fact that the Inuit have fifty different words for snow), so he and his companions sweated through the day in their reindeer-fur coats and boots as they pulled the sleds – the dogs they would normally have used were ill – and each morning woke in their igloos to find their clothes frozen solid. It was impossible, he told Marie, to express the joy with which they would greet the sight of a distant igloo after a march sometimes of more than twenty-four hours through -45° Celsius cold, and 'how comfortable and beautiful it seems when one enters into these dirty, narrow spaces, at the appearance of which I at first turned away in horror'.

He blushed, he wrote, 'to remember that during our meal tonight I thought about how good a pudding with plum sauce would taste. But you have no idea what an effect privations and hunger, real hunger, have on a person … the contrast is almost unbelievable when I remember that a year ago I was in society and observed all the rules of good taste, and tonight I sit in this snow hut with Wilhelm and an Eskimo, eating a piece of raw, frozen seal meat which first had to be hacked up with an axe, and greedily gulping my coffee.'

Gradually his ideas of superiority, as a European, began to wear away. The condescending parent–child relationship he had assumed only a few months earlier was replaced by a growing sense that he and his Inuit friends were equals. 'I am now a true Eskimo,' he exulted on that first expedition in December. 'I live as

they do, hunt with them, and belong to the men of Anarnitung.'
When Oxaitung, one of his guides, harpooned two seals, they
were divided between the settlement families amid celebration. 'Is
it not a beautiful custom among these savages [*wildun*] that they
bear all deprivations in common, and are also at their happiest
best – eating and drinking – when someone has brought back
booty from the hunt? I often ask myself what advantages our
"good society" possesses over that of the "savages" and find,
the more I see of their customs, that we have no right to look
down upon them. We have no right to blame them for their forms
and superstitions which may seem ridiculous to us. We "highly
educated" people, relatively speaking, are much worse.'

It was as if he had entered an icy Garden of Eden and he
was enthralled as much by the way he rose to meet his challenge
as by the challenge itself. Many years later, recalling this first
expedition of his career, he remembered 'days of the most joyful
feeling of freedom, of self-reliance: ready to meet the dangers of
the ice, sea, and wild animals; on the alert to meet and overcome
difficulties; no human being there to hinder or help'. No human
being, of course, except his servant, Wilhelm Weike, and his Inuit
guides and companions. Class and education were tighter chains
from which to escape than 'civilisation'.

Weike, who also kept a diary of his time in Canada, was less
interested in the ethnographic and scientific endeavours that so
fascinated Boas but arguably interacted more directly with their
hosts and, by having an Inuit girlfriend, was better integrated

into the society Boas observed as an outsider. Emotionally and culturally, Boas still held himself apart, writing constantly to Marie and his family and spending his evenings in the igloo reading Immanuel Kant in the light from an oil lamp improvised from a butter tin. Thoughts of Marie and their future together sustained him. 'And you, dear girl, will always help me. If my strength should weaken, you will give me renewed strength – just as you give me new strength here.' He held her farewell words close to his heart: '*Vorwärts* [Onwards], I wait for you!'

Boas may have seen his time in Canada as an adventure as well as a scientific expedition but, from the Inuits' point of view, the winter he stayed with them was notably more difficult than usual and they associated him with their hardship. Their sledge dogs were plagued by disease, so hunters and traders had to pull their sleds themselves, seals were scarce and the temperatures were unusually cold. For Boas, this meant his ink froze and he was forced to take notes and write his 'chicken scratchings' (as he called his tiny, almost illegible handwriting) to Marie in pencil; for his companions, it meant even more dangerous efforts to move around and find food and deaths that might otherwise have been avoided. Without fully comprehending it, the uninvited guest was a burden to his hosts.

Tragically, diphtheria (previously unknown there) and pneumonia also stalked the Inuit that winter and, although the Doctora'dluk had medicines to relieve fever and pain, not until the advent of penicillin in 1928 could these deadly infections, passed on by contact with the 'civilised' world, effectively be cured. 'Many Eskimo blamed me for it [the diseases], as it really seems as though sickness and death follow my footsteps. If I were superstitious, I really would believe that my presence brought misfortune to the Eskimo! Many are supposed to have said that they did not wish to see me in their iglu [*sic*] again.' Bitter hostility to Boas bubbled up; people were forbidden by the shaman from allowing him into their huts or igloos and from lending him dogs. Boas noted the distrust but remained resolutely positive, determined that this ill-will towards him 'should not be allowed to prevail'.

Not until March, having threatened to withhold trade of his valuable ammunition from Napekin, the shaman who had accused him of bringing bad luck, was Boas able to repair his relations with his unwilling hosts. At last healthy dogs were found and, in May, Boas, Weike and a guide, Sanguja, set off to chart Davis Strait, the central geographical aim of the expedition. It was a terrible journey: soft, heavy spring snow fell, the thawing sea ice was treacherous, they suffered from snow blindness and toothache, and provisions were so scarce they thought they would have to shoot and eat the dogs. 'I do not think I shall ever in my life forget horrible Home Bay!' exclaimed Boas. He and Weike began to long for home.

Having packed up their kit and notes, they waited for a ship in the tiny settlement of Idjuniving for several weeks. 'With longing I picture the time when nothing shall separate us,' Boas wrote to

Boas dressed up as an Inuit to pose for studio photos to illustrate The Central Eskimo. *He is wearing a seal-fur parka, an Inuit word he introduced to the rest of the world.*

Marie. Not until late August 1884 were he and Weike able to embark on a whaler for New York via Newfoundland. Having borrowed clothes from the captain – they had only their caribou-hide Inuit outfits – they landed in New York on 21 September, to be greeted by various members of the Jacobi and Krackowizer clans, and Boas raced upstate to Lake George, where Marie was waiting for him. While they announced their engagement officially then (at Boas's request, the announcement was backdated to 30 May 1883, the date on which Marie had accepted his epistolary proposal), it would be nearly a decade from when they met before her family considered him stable enough, financially and professionally, to marry her.

It took Boas four years to publish his account of his expedition to the Cumberland Sound. Although his delayed arrival in Kekerten, the uncharacteristically bad weather and the diseased dogs had hindered him from achieving all his aims, the work he had done there was important and admirable. *The Central Eskimo* (1888) contained alongside his geographical observations lists of Inuit tribes, their trades and manufactures, means of transport and navigation, igloo-building techniques, occupations and amusements, songs and poetry. As objectively as he could, he described the ivory goggles worn to protect the eyes from snow glare, the technology behind harpoons and kayaks, the kindness with which children were treated, the possibility that women washed their newborn babies in urine, the exchanging of wives as a sign of friendship, the tattooed faces of the Inuit women.

For Boas, the ethnographic research he'd done almost as an afterthought was what made his stay there meaningful: 'My work among the Eskimo satisfies me more than my travels.' Immersion in Inuit culture, living as his hosts did, had given him a new understanding of both 'savage' culture and so-called civilisation alongside a powerful sense of mission: 'I believe that one can be really happy only as a member of humanity as a whole, if one works with all one's energy together with the masses towards his goals.' He knew that he owed his life to his Inuit hosts, whose wisdom and experience had ensured his survival on more than one occasion. His time in the Arctic with them helped shape the egalitarian credo that would remain with him throughout his life and mark his long career. 'I had seen that they enjoy life, and a hard one, as we do; that nature is also beautiful to them; that feelings of friendship also root in the Eskimo heart; that, although the character of their life is so rude [*sic*] as compared to civilised life, the Eskimo is a man as we are; that his feelings, his virtues and his shortcomings are based in human nature, like ours.'

Boas hoped to remain in the United States after his year in Canada. Fearful of German restrictions that required job applicants to declare their religion, as well as a mandatory military draft, he saw the United States as a place in which he could 'further those ideas for which I live'. In addition, Marie was living in New York with her parents, friends of Uncle Jacobi's. At first, though, he returned to his own parents in Europe to determine whether his career lay in physics, geography, linguistics or folklore. He received the academic qualification, the habilitation, to which his dissertation on Baffin Island entitled him, but failed to raise the funds to return to the Arctic and continue his research. Working at the Royal Ethnological Museum in Berlin with the ethnographer Adolf Bastian, who believed that all humans had the same capacity to think and feel and that differences between cultures were the products of historical accident, tugged Boas ever closer towards a career in what today is known as anthropology.

After a year, he set sail for the United States to rejoin Marie and embark on life in a new world. The years that followed were hard. Still occasionally supported by his father and uncle, still considered too poor to be a suitable husband, struggling to learn English fluently, Boas pursued his fieldwork in British Columbia (in which he had developed an interest in Berlin) in parallel with what writing and museum jobs he could find, working as a junior editor at *Science* magazine. At last, in 1887, he and Marie were able to marry.

It was while doing field research among Native American tribes in British Columbia during this period that Boas collected skulls, sometimes buying them, sometimes sneaking into burial grounds by night, and sold them for $3 to $5 each to collectors and curators to fund his expeditions. It was, he confessed to his diary, 'most unpleasant work to steal bones from a grave, but what is the use, someone has to do it ... I dreamed of skulls and bones all last night. I dislike very much working with this stuff.'

Craniologists in the nineteenth century used skull size to theorise about racial types, seeing anatomy as the key to understanding the differences between peoples. At the start of the century, the influential physician and comparative anatomist Johann Friedrich Blumenbach had separated races into American, Caucasian, Ethiopian, Malay and Mongolian. Samuel Morton, a Quaker doctor and natural scientist from Philadelphia, who died in 1851, went further, arguing that God had created not a single human race but multiple races (polygenism), of which Caucasians were the superior type. The larger the skull and thus the brain, he proposed, the more intelligent the man. To compare 'the characters of the different races of men', he built up a collection of a thousand human skulls from all over the world and all periods of history, which included those of cannibals and criminals, prostitutes, soldiers, slaves and sadhus; what he hoped to prove was that some people, ineradicably and from birth, were inferior to others.

The parallel pseudo-science of phrenology, in which personality was divined from the shape of the skull, became at this time something of a fashionable amateur hobby. Could criminals and idiots, in the parlance of the day, be identified just by the shape or size of their skulls? When Boas bought skulls from a pair of phrenologist brothers near Vancouver in the 1880s, he managed to refrain 'from saying anything about the nonsense of phrenology. In the course of years I have acquired the curious habit of listening to all manner of opinions without agreeing or opposing.'

Despite his distaste for collecting and selling skulls, measuring them would become a leitmotif of Boas's career. His research with anthropometric measurements developed into a way of proving not that skull shape and size defined intelligence or any sort of innate superiority but that the physical appearance of skulls changed as living conditions and culture did. The defining classificatory system of humans was increasingly seen to be language – what went on in the brain, rather than what size it was – although in a neat twist Boas was later able to use Morton's collection to support his theory, generally accepted today, that changes in environment rather than inherited racial differences were responsible for divergence in skull shapes and sizes over generations.

In 1889, freshly appointed as head of the new department of anthropology at the recently founded Clark University* in Worcester, Massachusetts, Boas embarked on a study of the growth of local schoolchildren, probably inspired by the 1870s survey by one of his Berlin mentors, the physical anthropologist Rudolf Virchow. Worcester's population was expanding by as many as 10,000 people every year in the late nineteenth century, making it an ideal place to research whether immigration changed immigrants. The *Worcester Telegram* was appalled: it described the thirty-three-year-old Boas as 'a perverted old lecher' pawing the 'tender bodies' of Worcester schoolchildren, and suspiciously mentioned the sabre cuts and 'slashes over his eye, on his nose, and on one cheek'. But Boas's time at Clark was short. After measuring the heads of 17,000 children, he resigned in 1892, in protest against alleged infringements of academic freedom and perhaps partly because of this local hostility to his studies.

From Worcester Boas went to Chicago to assist the archaeologist and anthropologist Frederic Ward Putnam in assembling the ethnological and archaeological displays at the 1893 Chicago World's Fair, also called the Columbian Exposition because it celebrated 400 years since Columbus had arrived in America. Alongside full-scale reproductions of Columbus's three ships, Putnam, Boas and their team created an anthropology hall displaying Inuit artefacts and a

---

* A graduate-only university, Clark was at this time haphazardly run by Granville Hall, the first person in the US to receive a doctorate in psychology, and Jonas Gilman Clark, who founded the university with the profits he'd made in the hardware business.

Native American cliff dwelling. Their primitivism was intended to contrast with and demonstrate the achievements of European Americans over the past four centuries.

Boas persuaded fourteen Kwakwaka'wakw people from British Columbia to live in a reconstructed village for the duration of the fair as 'living specimens'. Today it is impossible not to see them and the Inuit people also performing for the visitors to the World's Fair as exhibits in a zoo – indeed, Ota Benga, a Mbuti man or pygmy bought from slave traders in Central Africa for a pound of salt and a bolt of cloth, was in 1904 exhibited in the Monkey House of the Bronx Zoo as a demonstration of the 'missing link' between primates and men. In German these displays of 'savages' were called *Volkerschauen*, or people shows; in 1880, a group of Inuit brought to Europe for such an exhibition all died of smallpox. Dahomey warriors, men as well as women, were also on show at the Chicago Fair.

In the end, only a small number of visitors were distracted away from the first Ferris wheel, Eadweard Muybridge's zoopraxiscope and Juicy Fruit chewing gum, all available in other areas of the Fair, to spend much time either with the 'living specimens' or in the eight rooms of the anthropology hall, of which Boas had been in sole charge – an anthropometric display that included a working laboratory conducting live research, measuring visitors' skulls in the name of science.

The Fair may have been a disappointment for Boas but his connection with Putnam would prove a saving grace. When he found himself unemployed again after reinstalling the (non-human) ethnographic displays in the new Field Museum in Chicago, Putnam offered him a job in New York at the American Museum of Natural History. Boas continued his work there, formulating the difference between displaying objects deductively (comparing them and drawing conclusions by analogy) and inductively (tracing the full history of each item within its own context). Thus, instead of stone knives from various places and times being on show together, the knives of a specific people would be displayed as part of a scheme about those people and how they lived. This scientific particularism would become a hallmark of Boas's academic work, like that of his intellectual contemporary, the psychologist William James, challenging the generalism of the previous generation.

From the museum he moved up Central Park to Columbia University, as a lecturer in 1896 and three years later a professor in its newly created department of anthropology, heading the PhD programme; there were only two other professors of anthropology in the country. (He suspected, but never found out for certain, that Uncle Jacobi had initially offered to underwrite his salary.) From then on, over the next four decades, the philosophy he had developed on Baffin Island would be the principle guiding the growth in the United States of the developing academic discipline of anthropology. As a young professor though, even with his laudable intentions, he sometimes found it hard to square his ideals with the practicalities of his colleagues' and contemporaries' views. Alongside his work at Columbia, he remained involved with the Natural History Museum for the next few years, collecting, cataloguing and curating objects for display. An incident he later hoped to forget illuminates the moral murkiness that swirled around his chosen field.

In 1897, while Boas was away in British Columbia, an Arctic explorer, Rear Admiral Robert Peary, brought six Inuit Greenlanders to the museum, half, it seems, as cultural informants, half as objects of study. They had been 'invited' to New York by Peary, and told they would be going home, but their status there had not been made clear; no plans had been made for their care. Three adults and one child died soon after their arrival from tuberculosis, a Western disease to which (as Boas knew only too well) the Inuit were especially susceptible. Minik, the seven-year-old son of one of the dead men, Qisuk, was adopted by the museum's chief curator, William Wallace. He pleaded for an Inuit burial for his father and, in 1898, although Wallace, Boas and the other staff at the museum knew that his father's body had already been dismembered and preserved in the name of scientific research, they created for Minik the ritual burial of a coffin he did not know was filled with rocks and built a sacred burial cairn with him in the museum's courtyard. Boas, who insisted the sham ceremony was only ever an attempt to comfort Minik, spoke a few words of respect and farewell for Qisuk. No one told Minik his father's skeleton would be on display inside the museum, labelled 'An Eskimo'.* It would take Boas the rest of his career to lay the foundations for a discipline in which this kind of objectification would no longer be permitted.

---

* Aged fifteen, Minik discovered that his father's bones were on display in the museum and was refused permission to take them back to Greenland. He returned to Greenland, essentially a foreigner, where the Inuit took him back and taught him the language and skills he had forgotten, and he became a hunter and guide. In 1916, he went back to the US, where he died two years later of influenza, in his late twenties. Not until 1993 were the bodies of Qisuk and three of his companions sent home for burial.

CHAPTER TWO

The Mentors

Alfred Haddon and William Rivers
in the Torres Strait, 1898

———•———

O ne of the sepia photographs from the University of Cambridge's ethnographic expedition to the Torres Strait in 1898 shows five scientists bearded and barefoot, looking almost like surfers, with rolled-up sleeves and loose khaki trousers, slightly eccentric broad-brimmed hats and deep tans. Gone are the collars, ties, studs, boots, black silk gowns and hoods that made up everyday dress for a Victorian academic: their contemporaries in college would hardly have recognised them.

Alfred Cort Haddon, the man sitting at the centre of the group with the dark crew cut and full beard, began his career as a zoologist and marine biologist. Ten years earlier, he had gone to study molluscs on the coral reefs of Papua New Guinea (then British New Guinea) and the islands of the Torres Strait, which separates it from the northern tip of Queensland, Australia. He left an account of his days living in a house in

Barefoot academics: this small team, with Alfred Haddon seated and William Rivers on the left, were the University of Cambridge's first ethnographic field workers.

Mabuiag, then known as Jervis Island but for which Haddon used the local name. Rising at six, he dressed ('such as it is, trowsers, flannel shirt and belt') and had a cup of cocoa ('made by my boy') and some bread before setting off barefoot to his makeshift laboratory half a mile down the beach. He spent the day capturing molluscs (invertebrates ranging from snails to octopuses), sorting, preserving and observing his catch and making notes. Like a sailor on a long trip, he drank water and lime juice and ate a mixture of tinned fruit, rice, fish (tinned or fresh) and very often dugong, or sea cow ('a swell dinner consists of dugong soup, dugong fritters, dugong steak, potatoes, rice and peaches'). Scrupulously, he took his daily dose of the malaria preventive, quinine.

He read after dinner, but only for a short time as the prevailing wind tended to blow his lamp out, and usually ended up with the 'missionary camp folk' at evening prayers. 'Imagine a blazing wood fire beneath a grove of young coconut palms', surrounded by local worshippers crouching on their haunches clad 'I am sorry to say in usually very grimy costumes'; those who looked best, he thought, were the young men in their traditional loincloths. The service was conducted in the local language, Lifu, by the schoolmaster or a missionary and began with a hymn. How strange, Haddon observed, 'does a familiar tune sound when wedded to native words and sung with native accent and emphasis, and original variations ... I was much amused when at the close of the service on my first evening here the conductor turned round to me and, just as if we had come out of church in England, politely remarked on the pleasantness of the weather.'

Afterwards he would return to his lodgings, where local friends would drop in, sometimes up to a dozen, 'and I endeavour to find out all I can about what they did "before white man he come – no missionary – no nothing". We have very pleasant times together, laughing and talking.' But all too often, he found, though the young men were happy to show him dances and describe their costumes and rituals, they said only the old men could really remember how things used to be. 'About half past nine or ten o'clock we say *Yawa* [goodnight].'

This 'yarning' with the 'savages',* learning about their changing society and material culture, made it clear to him that the islanders' uniqueness was being eroded by prolonged contact with Europeans, particularly the Christian missionaries who had been active in the region since the 1870s and who systematically destroyed any objects used in traditional ceremonies, seeing them as pagan. Following the influential American scholar Lewis Henry Morgan, Haddon believed that human society was progressing from primitive to civilised. He saw his friends (his word) in the Torres Strait as remnants of earlier cultures 'persisting in the fag-ends of continents', like human mudfish. This was no insult: coming from a zoologist, it meant they were utterly fascinating and merited urgent study.

When he returned to England, Haddon declared that the molluscs could wait: what was far more important was that the vanishing art and customs of these islands should be recorded before they disappeared altogether. He accepted a position at Cambridge as an ethnologist, which came with an even smaller income than other similar posts. His mentor, the biologist and anatomist Thomas Huxley (grandfather of the novelist Aldous), warned him of the difficulty of coming at bread, let alone butter, as an anthropologist, but Haddon's wife, Fanny, remarked comfortingly, 'You might as well starve as an anthropologist

---

* Haddon defined the noun savage as a 'dweller in the woods, i.e. backward people driven into out-of-the-way places' and the adjective as 'containing all those bad qualities which have been produced by contact with stronger peoples'.

as a zoologist.' 'I know of no department of natural science more likely to reward a man who goes into it thoroughly than anthropology,' Huxley observed to Haddon. 'It is one of those branches of inquiry which brings one into contact with the great problems of humanity in every direction.'

Within a couple of years, Haddon had produced a monograph on the *Decorative Art of New Guinea* but what he really hoped to bring about was 'a general work on anthropology', to be amassed in the islands by a group of specialists. It was Haddon, apparently, who introduced the naturalists' term 'fieldwork' into anthropology. 'A proper anthropologist' of the kind he needed would be a 'linguist, artist, musician and have extensive knowledge of natural and mechanical science etc.'. As his student, colleague and biographer Alison Quiggin observed, he should also have 'a gift for friendship ... regardless of accidents of rank or wealth'.

The expedition he proposed was the first of its kind in England. The grandfathers of the emergent discipline, Sir Edward Burnett Tylor at Oxford and Sir James Frazer at Trinity College, Cambridge, had done almost no fieldwork, relying instead on the ethnographic reports of far-flung missionaries and colonial officials; their work was more literary than scientific. In the universities of England, anthropologists had learned about distant and exotic groups from books. In the United States, as Franz Boas was realising, aspiring anthropologists and their predecessors had a ready-made subject matter, Native Americans, their very own 'primitives' to study. Haddon and his companions in turn discovered that the Empire, a map of the world two thirds pink, gave them astonishing access to the indigenous peoples they wanted to observe. In the 1890s, British social scientists realised to their delight that they could work almost anywhere on the globe and be assured of administrative support and, possibly, a spot of cricket.

Given that Haddon hoped to apply systematic methods of field observation to a discipline that until now had reached its theories and conclusions from the comfort of armchairs, perhaps it is no surprise that he approached practical people, doctors, biologists and linguists, before theoretical anthropologists. One of the first people he invited to join his expedition was a Cambridge colleague, a teacher of psychology at St John's College, William Halse Rivers Rivers (it is unclear whether the double Rivers was a mistake when his father registered his birth but he stuck by it). Rivers declined and Haddon recruited a young photographer, Antony Wilkin, and a linguist specialising in Melanesian languages, Sidney Ray. Charles Seligman, at this stage a physician and pathologist with an interest in tropical diseases but later, as a result of this journey, a noted professor of ethnology at the University of London, also joined the team. Then two of Rivers' former students and collaborators, Charles Myers, a psychologist and musician who was already a member of the Anthropological Institute, and William McDougall, a physician and psychologist, signed up; to Haddon's delight, Rivers decided to come too. Later

Haddon would say that he considered his greatest professional achievement to have been diverting Rivers from psychology to anthropology.

In 1898, William Rivers was thirty-four years old, tall, shy and with a slightly monastic air, 'a gift for invisibility' that would stand him in good stead as an ethnographer. He'd trained at the University of London and Barts and served as a ship's surgeon in the navy, travelling all over the world, before retiring from the forces to devote himself to 'insanity', psychology and neurology. After studying at Heidelberg and Jena, he was working at London's Bethlem Royal Hospital when he was recruited by St John's. As a boy, typhoid had caused him to miss his final school year, and therefore his chance to sit the Cambridge exams, but he had finally arrived at his destination. For the rest of his life, his beautifully panelled rooms in college, in 'an awful muddle, with books and papers and odds and ends of anthropological trophies all over the place', would be his permanent base.

Many of his students described their modest but inspirational teacher and mentor. When Rivers came through the door of his study, wrote one, Frederick Bartlett, later Cambridge's first professor of experimental psychology, 'somehow at once the room came alive ... You got a swift impression of straight, broad shoulders and a jutting chin, and at once of a tremendously alert mind.' His way of moving was quick and light, his way of thinking, according to another, 'quiet and alert, purposeful and unhesitating'. His power lay not in obvious charisma but rather in how he 'seemed to think not at all of himself, but cared so very much about you and your work'.

To the writer and critic Frank Swinnerton (a friend, not his student), he was physically 'an unimpressive figure, dressed with a sort of inconspicuous shabbiness. Always, across his face, there spread a wide smile ... I doubt whether he ever accepted any idea without considering it with the greatest care ... He gave the impression of being one of the most evenly tempered and kindly creatures I have ever met; and this with no suggestion of softness. Moreover he was not ridden by a theory and was not afraid of reality.'

Rivers' initial reluctance to embark on Haddon's anthropological adventure can perhaps be attributed to James Hunt, his maternal uncle. Hunt's day job was as a speech therapist and his patients included two great figures of Victorian children's literature, Charles Kingsley (*The Water-Babies*) and Charles Dodgson (as Lewis Carroll, *Alice in Wonderland*), a family friend, at his institute at Ore House near Hastings. His interest in speech had made him an evangelistic anthropologist, an early member and joint secretary of the Ethnological Society in the 1850s and 1860s, who ensured that the British Association for the Advancement of Science accepted anthropology as a discipline in 1866. He was also an active extreme polygenist and negrophobe who believed the human races descended from multiple sources (which would mean they were biologically unequal) and defended slavery on those grounds; even in the nineteenth century, his views were seen as abhorrent. Hunt died in 1869, when Rivers was six, and his brother-in-law, Rivers' father, took over his practice. When he was older, Rivers would turn down his uncle's anthropological library: he did not want it. Perhaps it was no accident that he also had a debilitating stammer that access to the family practice does not seem to have improved.

Assisted by Myers and McDougall (and echoing Boas's dissertation on the colour of water), Rivers' plan in the Torres Strait was to study the colour perception of the islanders by giving them visual tests; he was to continue these tests in Egypt several years later, as well as with an Inuit man from Labrador visiting England and with English schoolchildren. His conclusion was that 'civilised' and 'uncivilised' peoples had the same visual acuity but civilised people were more sensitive to the colour blue. Strangely, given the colour of their sea and sky, the Papuans had no word for blue.

While studying the islanders' physiology, Rivers became interested in the genetic connections between individuals and 'began to collect genealogies in order to ascertain how far aptitudes or disaptitudes ran in families'. Guided by the work of Lewis Henry Morgan, which placed kinship study rather than the development of religion at the heart of anthropology, he threw himself into an exploration of family structure, inventing for the purpose what he called the 'genealogical method' of recording terms for relatives, which permitted reconstruction of ages-old social structures – cultural fossils, as he would put it. Kinship, or systems of relationship, governed behaviour and regulated etiquette and as such was seen as the key to how society functioned and had originated; the words by which people referred to blood and marriage relations were intimately connected to long-standing social structure and customs. But the data he gathered was recalcitrant: he found no cross-cousin marriage, nor any definite exogamy, marrying out of social groupings. This made it impossible to trace the linear progress conceived by Morgan and his followers from primitive to civilised social structures.

Like Boas in the Arctic a decade earlier, during this trip Haddon, Rivers and their friends developed a lifelong fascination with string figures, or cat's cradle. These games feature in societies all over the world; no other game, historically, has been so widespread. They have a practical purpose – string is essential for hunting, fishing, weaving and medicine – but importantly were also used by shamans for divination, story-telling and entertainment. Louis Leakey, mentor to Jane Goodall and Dian Fossey, among others, was taught by Haddon; he credited his advice, 'You can travel anywhere with a smile and a piece of string', with saving his life on his arrival among hostile tribes in sub-Saharan Africa. Haddon's daughter Kathleen, a zoologist, would grow up to become one of the world's specialists in string figures.

It would become a truism of mid-twentieth-century anthropology that fieldwork was a gruelling, boring necessity but Rivers, Haddon and their companions found their time on the palm-fringed islands of the Torres Strait stimulating and enjoyable. Despite their fears about the ferocity of the islanders, who were rumoured to be cannibals with a taste for Europeans,* they realised, in Haddon's words, that instead the natives had suffered from 'the malpractices of the white man'. The team were nearly shipwrecked on the Great Barrier Reef and, at one point, Rivers' and Sidney Ray's legs were so badly sunburned it took them more than a week to recover but they loved the work they were doing, understanding its importance, and gelled as a group, appearing relaxed and happy in the photographs taken by Wilkin.

Rivers' health, fragile since his teenage bout with typhoid, was always improved by travel; his mental state benefited as much from the professional camaraderie in the Torres Strait as from the tropical breezes. He had been anxious and discontented before this trip but returned to Cambridge refreshed. Although he carried on with his medical and psychological research, he and Haddon continued to promote 'our Cinderella science'. As a fellow at Christ's College until he retired in 1926, Haddon mentored a generation of anthropologists including Alfred Radcliffe-Brown and Bronislaw Malinowski and was a valued friend and colleague of Edvard Westermarck (see Chapter 3). His inspiring brand of anthropology, wrote Quiggin, 'might be called Philanthropology ... his service to humanity was to show that "the proper study of Mankind" is to discover Man as a human being, whatever the texture of his hair, the colour of his skin or the shape of his skull'.

Almost immediately, his imagination sparked by his time

---

* Some inhabitants of the more remote areas of New Guinea did practise head-hunting and ritual cannibalism through the twentieth century.

in the field with Haddon, Rivers began to plan another trip to study kinship and genealogy. In 1902, he went to southern India to observe a group called the Toda, who were small and discrete enough for him to gather data on every member. Staying in a hotel in the little hill station of Ootamacund, in the Nilgiri Hills where the Toda lived, he researched the book that would, as he put it, 'make ethnology a science'. It was here, in India, that Rivers' 'scientific detachment and experimental rigour, wholly different from the spirit of classical scholarship' pervading the work of his older Cambridge colleague, Sir James Frazer, would irrevocably shape the progress of British anthropology.

Although Rivers advocated cultural immersion for field workers, he stayed in an Ootamacund hotel rather than one of these thatched Toda houses, photographed c. 1900.

The Toda were perfect subjects for Rivers. A small, patriarchal society, they were dairy farmers whose cattle formed the basis for their unique ritual religion, neither Hindu nor Muslim. Women had more than one husband and sometimes men had more than one wife – but although their society was promiscuous, marriage was strictly regulated. When a woman married a man, she automatically became his brother's wife, too, and her children belonged to the household in common. Wives could be exchanged for buffalo. There was no word for adultery in the Toda language; it was the man jealous of his wife's freedom who was considered immoral. Female infanticide was common, which was one reason polyandry was practised – women were scarcer than men.

Ethnographic surveys had long been a feature of colonial life and Rivers was by no means the first researcher the Toda had encountered. What was special about his work with them from an anthropological point of view were his methods: 'his insistence on scientific procedure, his delight in scientific analysis, and his facility in adapting scientific methods to novel experimental conditions'. He apologised in the preface to *The Todas* (1906) for the 'minute detail' into which he went describing his technique, but 'it has been my object in writing this book to make it, not merely a record of the customs and beliefs of a people, but also a demonstration of anthropological method'. Clear distinctions between descriptions of custom and belief and reliable, meticulous collection and recording of information on one hand and any 'theoretical conclusions drawn' on the other would mean that if his processes were followed, one researcher's work could be accurately consulted by another at a later point; indeed, Rivers' Toda work was used profitably by anthropologists for decades afterwards.

Looking back on his time in India, Rivers concluded that a 'typical' – perhaps more accurately ideal – piece of intensive fieldwork would be 'one in which the worker lives for a year or more among a community of perhaps four or five hundred people and studies every detail of their life and culture; in which he comes to know every member of the community personally; in which he is not content with generalised information, but studies every feature of life and custom in concrete detail and by means of the vernacular language'. He hadn't followed his own advice, having used interpreters and stayed in the comfort of an Ooty hotel for only six months, but these were the lessons he had learned. His words would become the statement of intent for all future British anthropologists working in the field.

As for the Toda in particular, Rivers found that they, like the Torres Strait islanders, did not fit neatly into his Morgan-inspired kinship theory. All he could conclude was that neither 'clan' (primitive and, according to the theory, promiscuous) nor 'family' (civilised) was a recognisable Toda institution and 'their state of social evolution' was 'intermediate'. It was a problem he would return to grapple with.

Towards the end of his time in the Nilgiris, like Boas with the Inuit, Rivers came up against local resistance to his enquiries.

Three of his guides and interpreters suffered bad luck or tragedy – one man's wife died after he brought Rivers to watch the most sacred of the Toda ceremonies – and the Toda priests concluded that their gods were angry with them for revealing their secrets to an outsider. Rivers, who seems to have felt no compunction about trying to persuade his informers to be less 'reticent', seems also to have felt no sense of abashment or responsibility about this; despite the 'sympathetic understanding' for which he was admired, presumably he saw Toda beliefs as nothing more than primitive superstition to be observed but not respected.

Rivers was careful to view his subjects with what he believed was scientific objectivity but which to later readers still reveals layers of bias: 'The characteristic note in the demeanour of the people is given by their absolute belief in their own superiority over the surrounding races. They are grave and dignified, and yet thoroughly cheerful and well-disposed towards all. In their intercourse with Europeans, they now recognise the superior race so far as wealth and the command of physical and mental resources are concerned, but yet they are not in the slightest degree servile, and about many matters still believe that their ways are superior to ours, and, in spite of their natural politeness, could sometimes not refrain from showing their contempt for conduct which we are accustomed to look upon as an indication of a high level of morality.'

Alongside his anthropological work, back at Cambridge Rivers continued with his experiments in physiology and psychology. He was elected a fellow of St John's in 1902 and given a small laboratory. From 1903, every Friday, he tested nerve division, sensation and regeneration with his colleague and friend Dr Henry Head (the almost unbelievably stoical Head sitting with his eyes closed while Rivers operated delicately on the nerves in his forearm); from 1906, the first to use the double-blind procedure, he studied the effects on himself of alcohol and caffeine; he also experimented with the psychological effects of sleep deprivation. 'Perhaps no man,' commented his friend and Torres Strait companion Charles Seligman, 'has ever approached the investigation of the human mind by so many routes.'

In 1908, Rivers returned to the South Pacific. He had obtained funding from the Percy Sladen Trust* for himself and a colleague, Arthur Hocart, to study the people of the Solomon Islands, working primarily on Simba Island. (Rivers was in Fiji for a month; Hocart remained for three years.) The resulting two volumes, *History of Melanesian Society*, published in 1914, were a demonstration of the anthropological method Rivers had laid out in *The Todas*. At the end of the ethnographic study, he outlined the conclusions his work had helped him draw. Studying family

---

* Percy Sladen was an English biologist (specialising in starfish) who died in 1900 aged fifty; his widow set up the trust in his name to further scientific research. Christine Dureau, in 2014 (Hviding and Berg, eds), asked, 'How to write about ... [the] 1908 Percy Sladen Trust Expedition to the Solomon Islands? The mere title of their undertaking – an expedition – evokes a journey into a recently pacified area aboard a colonial mission yacht in order to investigate a social evolutionist project.' And yet, she adds, for their time Rivers and his companions were humane, opposed to Eurocentrism and colonial insensitivity, and pioneers of empirical, participant-observation fieldwork.

relationships, or kinship, remained 'essential and fundamental' to any ethnological research; secret societies or religious cults were often repositories of ancient culture, forgotten elsewhere in a society's daily life; language and its structures were vital tools for the student of human culture. Finally, the overriding lesson he had learned was that all Melanesian culture and society was 'the direct outcome of the interaction between different and sometimes conflicting cultures'. Interaction and conflict produce culture; even small numbers of migrants can cause seismic change.

This was a dramatic transformation of his thinking. While he still subscribed to Morgan's belief that kinship was at the heart of anthropology, he now believed that change occurred in societies not because of a preordained progression from primitive to civilised, but as a consequence of the spread of people, culture, ideas and technology between societies. It meant that culture was constantly changing and no society was ever static; more importantly, there was no civilised 'end' towards which groups of people were inexorably moving. This theory, prevalent in German intellectual circles and also familiar to Boas, was known as diffusionism and Rivers would be wedded to it for the rest of his life.

He had also been greatly influenced by his friend and colleague, and possible student at St John's in the 1890s, Grafton Elliot Smith, the leading thinker on brain anatomy and evolution. Working in Cairo as an archaeological consultant in the early 1900s, Smith was the first person to X-ray mummies.* He believed, and Rivers believed with him, both that modern humans originated around the Mediterranean (which the archaeological evidence at that time supported) and that all human civilisation anywhere in the world ultimately derived from Ancient Egypt, from where it had radiated outwards through cultural diffusion. Despite Smith's undoubted scholarship and anatomical expertise, extreme diffusion (as it became known) was a kooky theory, even then; association with it would later allow Rivers' detractors a foothold from which to challenge him.

The other point for which *Melanesian Society* is remarkable is the directness with which Rivers catalogued the 'devastating effects of our "civilisation"' upon Melanesian culture and history. He speculated that not physical but 'psychological factors' – apathy, boredom and hopelessness, brought on by colonial contact and its attendant demons, venereal disease and alcohol abuse – were causing dramatically reduced birth rates in Melanesia. In his view, the people of Melanesia were committing 'racial suicide' because, after the arrival of Europeans and the desecration of their world and way of life, they no longer saw any point in living.

A year before publication, in 1913, Rivers had submitted a report to the Carnegie Institute about the necessity of anthropological investigation and the places that in his view most

---

* He found evidence of biparietal thinning in many Egyptian skulls, a condition also common in eighteenth-century European aristocrats, and speculated that it was caused in both by wearing heavy wigs or headdresses.

merited study. 'In many parts of the world the customs of savage and barbarous man are undergoing rapid and destructive change.' (I think that by 'savage and barbarous' he only really meant not 'civilised': he was as conscious as anyone of how savage and barbarous supposedly civilised men could be.)

There were, he said, two kinds of ethnographic enquiry: surveys, which had the benefit of being wide-ranging but were by definition superficial; and intensive study, in which an observer would live immersed in a society for a year or more, speaking its language. This was the ideal but it was impossible to do this kind of intensive work among people 'untouched by Western influence ... It is only a people already subject in some measure to the mollifying influence of the official and the missionary who will not fear, or be offended by, inquiry into their customs.' Perhaps he was thinking of the Toda, with their unhelpful concerns that he had brought a curse down on them. Between ten and thirty years after colonialisation had been effected was, he thought, the optimum moment for an ethnologist to arrive, though he added that 'measuring' natives and taking samples from them, for example snipping locks of hair, 'deeply offends' them.

It was vital that the observer be a trained anthropologist. Although colonial officials and missionaries might be well placed to collect information, they were untrained and their roles were in opposition to 'native ideas and custom': the official wants to subdue them, the missionary to eradicate and replace them. A sympathetic ethnologist, on the other hand, could gain the confidence of his

subjects and thus 'obtain objects [and information] ungrudgingly
... [and] everything he obtains will have an infinitely wider and
deeper meaning than anything which can be obtained by the
cursory visitor'.

Soon after this report and the publication of *Melanesian
Society*, the savagery and barbarism of the so-called civilised
world erupted in the catastrophe of World War I. In 1915, having
returned from another trip to the South Pacific, Rivers joined the
army as a doctor and was soon appointed a psychiatrist dealing with
shell-shocked soldiers. He worked for a year at Maghull Military
Hospital near Liverpool before being sent to Craiglockhart War
Hospital outside Edinburgh, where his wise and compassionate
treatment for 'war neuroses' made him beloved among his patients,
who included Siegfried Sassoon, Robert Graves and Wilfred
Owen.* Sassoon remembered Rivers' face illuminated by the
late-summer sun in their thrice-weekly meetings, his spectacles
pushed up on his forehead and his hands clasped over one knee,
gently persuading the devastated poet to believe 'he liked me and
he believed in me'. In 1918, Rivers was appointed psychologist
to the new Royal Flying Corps in Hampstead; aged fifty-three,
accompanying his patients, he became an enthusiastic passenger
on test flights of the notoriously unreliable Sopwith Camel.

With the onset of peace, a happier William Rivers returned
to Cambridge. Diffidence had been replaced by confidence and

---

* Rivers' wartime career as a psychologist and psychiatrist merits its own book. Most
memorably, his time at Craiglockhart lies at the heart of Pat Barker's *Regeneration* trilogy.

reticence by forthrightness. The enthusiasm for others that had always marked his teaching style became more open and more easily given. His work on the unconscious, as he sought to help his shell-shocked patients, had made him attempt to unravel his own tangled interior life. In *Instinct and the Unconscious* (1920), he revealed that for the past two years he had been 'attempting to penetrate' the mystery behind what he described as an inability to remember, his lack of visual memory from about the age of five.

What interested him was that he had very detailed memories of the lower floors of the house he lived in until that age but could recall nothing at all of the upper floor – where, as a Victorian child of the upper middle classes, he would have spent the majority of his time. For years he had assumed that his lack of visual memory was due to the fact that as he grew up, he had become more interested in the abstract; in this book, he had to conclude that something had happened to him on that floor, the memory of which was suppressed because it would have 'interfered with my comfort and happiness'. Though he recalled many childhood incidents that took place on the ground floor, 'no event of any kind which happened in the upper storey has ever come to my consciousness. Now and then, when in the half-waking, half-sleeping state, peculiarly favourable in my experience to the recovery of long-forgotten events, I have had the sense that something is there, lying very near emergence into consciousness. But I have not yet succeeded in penetrating the veil which separates me from all knowledge of my life in that upper storey.'

On a professional level, he began to integrate his two passions, psychology and ethnology, disciplines that until then he had viewed as separate. The realisations he had made in Melanesia about collective trauma were applicable to European society, too, still reeling after the violence and slaughter of the war, tormented by blood-soaked nightmares, insomnia, tics, convulsions and paralysis: people were people, wherever they were and whatever their culture. Now he understood that 'the thoughts and behaviour of any community were worthy of study as a means of understanding the psychology of mankind as a whole'.

The friendships he had made during the war, notably with Siegfried Sassoon, endured afterwards. Sassoon dated his own 'mental maturity' from the time he met Rivers, whose integrity, guidance and support he valued above any other. The novelist Arnold Bennett, who became a friend during this period through Sassoon, left a long pen portrait of Rivers, whom he described as 'a hero of the first order to many'. He described a man of deep and simple modesty, with spartan personal habits – not because he was depriving himself of anything, but because he needed nothing but the life of the mind. (Bennett added that although non-drinkers are 'always a worry', Rivers was an exception.)

Watching him with students, said Bennett, was especially touching because of the 'fallacious but charming equality which the elder established and maintained between the two [Rivers and a student] … the sweet and skilful wisdom', his refreshingly unconventional way of thinking and his 'laudable' ideas about tea and cake. 'It was less his universal knowledge that impressed me than his lovely gift of coordinating apparently unrelated facts. And it was less his gift of coordination that impressed me than the beauty, comeliness and justness of his attitude towards life ... A doctor of medicine, he had little belief in current therapeutics. He said, apropos of a recent indisposition: "I thought I'd better call in the magician, and he prescribed something or other. Anyhow, I got better." (All civilised society was a sort of South Sea Island to him.)'

Rivers' war work had pushed his politics, previously neutrally establishment, further to the left. In 1922, nominated by Harold Laski and Bertrand Russell, he accepted the Labour candidacy for the parliamentary seat of the University of London: 'To one whose life has been passed in scientific research and education the prospect of entering politics can be no light matter. But the times are so ominous, the outlook, both for our own country and the world, so black, that, if others think that I can be of service in political life, I cannot refuse.'

That same year, Rivers suffered a strangulated hernia, alone in the night in his rooms at St John's. He was still alive when he was found but in terrible pain and, despite an operation, there was nothing that could be done to save him. He died on Sunday 4 June 1922, aged fifty-eight. His friends and colleagues greeted his death with encomia. 'It is impossible to indicate what his death means to his many friends,' wrote Alfred Haddon, while Bertrand Russell praised a potential MP uniquely 'devoid of political ambition

... [and] exclusively actuated by public motives'. Until his own death Siegfried Sassoon would address poetry to Rivers, his father figure and confessor. 'The influence of his vivid personality will remain for all who knew him as one of the best things that have ever entered their lives,' said Frederick Bartlett.

Rivers' anthropological successors, though, were not always as generous. His devotion to diffusionism would be drowned out by the 'twin but discordant trumpetings of Bronislaw Malinowski and A. R. Radcliffe-Brown' as they championed their differing forms of functionalism (the anthropological theory that all aspects of a society serve a function and are necessary for its survival) from the mid 1920s onwards. His dependence on genealogies was derided by Malinowski as 'kinship algebra'; even his innovations in fieldwork were belittled as Malinowski declared that if Rivers was Rider Haggard, he himself would be Joseph Conrad. Across the Atlantic, Robert Lowie, one of Boas's students, criticised what he said was Rivers' belief that the 'natives' he studied were so overawed by white men that they willingly submitted to his requests and were ready to abandon all they knew to adapt to his culture; actually, Rivers' understanding of the Toda sense of superiority chimed very closely with Lowie's field experiences among Plains Indians of the American West, 'who, man to man, regard themselves in no way inferior to whites'.

Perhaps the best comparison, from the standpoint of the development of the discipline of anthropology, is with Franz Boas, only four years Rivers' senior. Boas's career lasted more than twenty years longer than Rivers' and, unlike Rivers, he devoted himself exclusively to anthropology; but, if the paternalistic Boas is the father of American anthropology, Rivers can surely claim to be one of the founding uncles of British anthropology. Quite apart from his contributions to psychology and physiology, he must be remembered as the man who made ethnology a rigorous scientific discipline, constructing what he called a coherent 'science of social psychology'. He was the leading figure in the replacement of religion with kinship as the central, defining study of anthropology; he lent his influential weight to the movement away from social evolutionism; but most importantly, according to Edmund Leach in 1968, it was 'because of Rivers that most British social anthropologists [began to] ... think of themselves as being engaged in a science rather than a literary exercise'.

In the months before his death, Rivers had given the presidential address to the Royal Anthropological Society, warning aspiring ethnographers not to forget the spirit of scientific endeavour that should illuminate all their work. 'There is, I think, a danger that the various topics before us may resemble the objects on the walls and in the cases of so many of our museums in being regarded merely as curiosities, each perhaps having some special interest of its own, and that in the wealth of the details of anthropology we may fail to recognise the threads which connect them into a coherent whole.' Kinship and diffusionism aside, his quest for a coherent whole, using fieldwork conducted in the spirit of rigorous scientific enquiry, was Rivers' great contribution to anthropology.

CHAPTER THREE

The Philosopher

Edvard Westermarck in Morocco, 1898

———•———

'As if by the wave of some magic wand I found myself suddenly transported to the East,' remembered Edvard Westermarck of his arrival in Morocco, with greater regard for his dreams of the exotic Orient than geographical accuracy. Waiting in vain in Gibraltar for a ship headed to the Moroccan port of Tetuan, the first stop in a projected round-the-world ethnographic tour, the thirty-five-year-old professor had impatiently embarked on a little steamer bound for sleepier Tangier instead. Hours later, he landed in another world. Veiled women and water-bearers with swollen goatskins on their backs gliding along the city's narrow, winding streets past windowless white-plastered walls and arched doorways, 'the stench of dirt and fat', the crescent-moon curve of the bay and the luxuriantly green mountain behind which lay the Atlantic: all these cried out romance and adventure. Eyes shining behind his neat gold-rimmed spectacles, Westermarck was enraptured.

It was the spring of 1898 and Westermarck, brought up and educated in Helsinki, had already found his way to the 'cathedral' of the British Museum's Reading Room, for him the beating heart of high culture and civilisation, 'the dwelling of the goddess of thought ... the storehouse of all the knowledge garnered through the centuries'. Here, he wrote, the reader – surely Westermarck himself – was no mere recipient of information but felt suffused with 'creative pride; he has a sense of being part of the whole, and enjoys something of the bliss of the god-inspired pantheist'. To Westermarck, a philosopher as well as an anthropologist, scholarship was akin to spirituality.

Westermarck's inspired labours in the Reading Room and in the nascent department of sociology at the neighbouring University of London had led to his first book, *The History of Human Marriage* (1891), seven years before he set foot in Tangier. With an introduction by no less a person than Alfred Russel Wallace (the eminent biologist and geographer who had arrived at the theory of evolution through natural selection independently of Charles Darwin four decades earlier), it placed Westermarck, then only twenty-eight, at the forefront of the emerging schools of British social science. The publication of this book and the positive reception it received were 'without doubt the most momentous happening in my life', wrote Westermarck, riffing on the old adage of Samuel Johnson's that marriage has many pains, but celibacy no pleasures. 'It has been said that marriage has many thorns, and celibacy no roses. For my own part I would say that marriage has brought me many roses – and bachelorhood no thorns.'

Following this success, like a monk abandoning the cloister for the worldlier existence of a mendicant friar, he resolved to study 'not only from books but from life'. At this stage in England there was no sense at all in the academic world that an anthropologist – and even this name for the discipline was a new and unfamiliar term to most academics – might need to set off beyond the college and library and plunge into the world of people to study them. Franz Boas had returned to Germany from the Arctic by this time, but William Rivers and Alfred Haddon had not yet embarked on their journey to the Torres Strait. Westermarck sat at his favourite desk beneath the cupola of the Reading Room and dreamed up an ethnographic expedition that would take in every corner of the globe. What he planned was nothing less than a comprehensive comparative study of all the primitive cultures of the world, with particular reference to 'the origin and development of moral ideas' – discovering how they led to the art, literature and philosophy of the Western culture he so prized.

As a philosophy student in Helsinki more than fifteen years earlier, he had proposed the thesis 'Does Civilisation Increase the Happiness of Mankind?' Although the adjudicator disagreed with him, he did award him the highest mark for his doctorate; and this question, alongside the young Edvard's affirmative answer, would lie at the heart of all his future work, both anthropological and philosophical. 'Why do the moral ideas in general vary so greatly in different individuals and among different peoples?' And, on the other hand, why is there in many cases such a wide agreement?' It was to investigate these questions that he conceived of his expedition, with Morocco as his first stop. In his pocket he bore 'a letter of introduction to an Englishman in Morocco' from someone (he could be charmingly vague) at the British Museum.

Westermarck was by no means the first Western traveller to Morocco, although in many ways the country was still as 'unadulterated' (his word) as any turn-of-the-century ethnographer could have hoped, and the prevailing European view of Moors, dating back to the sixteenth century, was still of bandits gripping metaphorical knives between gleaming teeth, keen to kidnap – and worse – all foreigners. The watercolourist Elizabeth Murray, who lived in Morocco in the 1840s, was persistently reminded of scenes from the *Arabian Nights* and the Bible during her time there and remarked upon the well-known hostility of Moors to Christians, as well as the 'filth' of the towns. Eighty years later, Edith Wharton was still very much in thrall to Orientalist tropes, marvelling at the romance of Morocco's deserts and mountains and its 'untamed' but 'picturesque' inhabitants: 'If one loses one's way in Morocco, civilisation vanishes as if it were a magic carpet rolled up by a Djinn.'

Scientific visitors had included the botanist Sir Joseph Hooker of Kew (Charles Darwin's closest friend), alongside his colleague John Ball, in 1878, and a medical missionary, Robert Kerr, who spent many years tending to Moroccans (as he put it) in both 'the palace and the hut'. The *Times* correspondent Walter Harris, buried in the English churchyard of St Andrew's in Tangier, published several authoritative accounts of his travels around Morocco in the 1880s and 1890s; the intrepid Robert Cunninghame Graham, later the first socialist MP and the first president of the Scottish Nationalist Party, went in search of the lost city of Taroudant disguised as a Turkish doctor.

All were beguiled by Morocco – 'Which of us has yet forgotten that first day when we set foot in Barbary?' enquired the opening line of the 1905 guide *Life in Morocco and Glimpses Beyond* – but perhaps the most charming account came from the explorer Joseph Thomson,* sent by the Royal Geographical Society to survey the country in the late 1880s after more ambitious expeditions to East Africa and the Niger river. On his first morning in Tangier, Thomson opened his window and knew himself transported from Europe to Africa: 'Before us a succession of whitewashed houses rose tier above tier, reflecting back the rays of the morning sun with blinding power.

'Minaret and dome rising from mosques relieved the somewhat monotonous aspect of the house-terraced slope, while the battlemented walls and frowning ramparts of the Kasbah or citadel, which overlooks the town, formed a broken outline projected sharply against the marvellous deep blue of the sky … Right beneath our window, on the other side of the four-feet

---

* His motto was said to be 'He who goes gently, goes safely; he who goes safely, goes far.' Thomson's Falls in Kenya and the Thomson's gazelle are named for him.

broad street, we peep into a native school, and see small boys squatting on the floor round a venerable be-turbanned *Taleb* [scribe] or teacher, while, with blatant discord and much swaying of bodies, they repeat vociferously texts from the Koran inscribed on wooden boards, thus learning at once the precepts of their faith and a smattering of classical Arabic.

'Farther off we hear some wildly attractive though barbaric music, with shouting, singing, and firing of guns, which we conclude must proceed from a wedding-procession ... Tantalising we find it in the extreme to crane our necks and look down on the *haik*-draped Moors or blanket-shrouded figures of the race. We are too high up to see their gazelle-like eyes, and can only feel in imagination their irresistible glances.

'As we bathe in the wealth of morning light, see overhead the clear blue sky, and feel our faces fanned by the fresh breeze, we cannot but picture the environment of our friends at home, who are beginning to struggle reluctantly out of bed, and shiver in their cold rooms, while they curse the sleet, the rain, the east winds, the fog, and their luck generally. We generously wish we could send them some Moorish weather, or mix for them a pot of our local colour.'

Westermarck was in wholehearted accord with Thomson's view of the local colour. After a few days' rest in Tangier, he continued on to Tetuan along a narrow coastal path, a small caravan of two horses and a mule, with the ancient Moroccan soldier appointed to accompany him through the kingdom at the head and a donkey boy bringing up the rear in charge of his luggage. For the last stage of the journey, the whitewashed walls of Tetuan gleamed ahead of them, gladdening 'my eyes like an apple-tree in full blossom'. Morocco in the springtime was an enchanted place, he remembered of this introduction to the country that would become an adopted home, 'with its colours, its scents, its sweet air and sunshine, its teeming life, and the song of its nightingales – this, indeed, belongs to those festive hours of life that fill the human soul with bliss'.

Following three intensive weeks learning Arabic with a tutor in Tetuan, Westermarck set off into the interior, heading for Fez. There were no roads in rural nineteenth-century Morocco, nor places to stay; he would have to travel across country, camping every night. It recalled the summers of his student days, tramping across the Scandinavian highlands, staying in crofters' cottages. Joining forces with an English tourist with whom he could share an interpreter, the party additionally consisted of the requisite official soldier, four Moroccan servants and the Englishman's valet. They had four tents, three horses, seven mules and a donkey and they took equipment and provisions for several weeks.

For someone who confessed that as a young man he 'could not feel happy without seeing a travelling-trunk in my room', the journey to Fez was, as Westermarck wrote long afterwards, a time of 'indescribable delight'. Despite riding beneath the burning sun, often for twelve hours a day, despite the poor food and dirty water, despite the flying, crawling and jumping insects that menaced him in his tent at night, despite being robbed of everything he possessed, including three months' worth of field notes just before his departure from Morocco, he was enchanted. He felt as if he had been carried back in time a thousand years, fancying himself 'a wayfarer in Europe at a period when made roads and bridges were unknown and the traveller had to make his own way as best he could'. What was astonishing was that this complete foreignness was only a few days' journey from London because it seemed like many centuries past as well as thousands of miles distant.

And even better than this romance, to which his soul was acutely attuned, was being among people 'who had kept their ancient belief in magic powers and mystic spirits that dwell in every mountain chasm, every spring and river, and creep out of their hiding-places when twilight falls and terrify the hearts of the children of men'. When Westermarck left Morocco after his first visit, it was as if he had fallen in love: he had found his life's subject. He wrote excitedly to his parents in Helsinki of 'all those genial Moors and interesting matters which become more and more interesting every day the more one begins to feel at home'.

He had learned, too, that unlike botany, zoology or geology, ethnological research required complete immersion in a culture – starting with fluency in its language, since direct communication with the 'natives', he'd realised, was irreplaceable – as well as long-term time and patience, the willingness (as Franz Boas would later tell Margaret Mead) to waste time. 'I had originally only looked upon Morocco as one stage on my way, and intended afterwards to go on to Ceylon, the South Sea Islands, and Heaven knows where. But that journey was never to come off.' Instead he returned to Tangier within months.

For the next thirty years, when he wasn't teaching in London or Finland, Morocco would be the focus of Westermarck's research. Although over the years he spent increasing lengths of time in congenial Tangier, where in the 1920s he bought a cottage on the Old Mountain from his friend the Swedish consul, usually he could be found accompanied by his 'invaluable' guide, Sîdi Abdsslam, and an armed retinue of eight, travelling across country from remote villages of the High Atlas as a guest of the Sultan to the medieval alleyways of Fez. Always in his luggage he carried a rifle and his two-volume edition of Sir James Frazer's *The Golden Bough*; shedding his three-piece suit, he adopted a crimson tasselled fez and flowing jellaba over his riding breeches. Sometimes he stayed somewhere for weeks or months; at other times he moved rapidly across country, breaking camp after a day or two. His longest time in Morocco was a period of twenty-six months in 1901–2.*

---

* Lady Grove, a daughter of the ethnographic collector General Pitt Rivers, was in Morocco at the same time and published her account of a long camping holiday there in 1902. Although she bemoaned the 'evil-smelling alleys' of Tangier and the 'wretched touts' pestering tourists at every turn in their efforts to sell them daggers, she concluded complacently that a winter there was 'for the most part, very much like a winter anywhere in a Southern watering-place'. There were games of rounders, tea parties, picnics, pig-sticking and paper chases, all the pleasures of colonial life; and, for a touch of the exotic, one rode to dinner parties and dances on donkeys.

He credited the youthful summers he had spent hiking over the mountains and fields of Finland and Norway with freeing him from a sense of awkwardness around strangers – particularly, as he put it with scrupulous precision, those of a 'lower social rank'. The ease he learned in those months of democratic ramblings living alongside peasants and crofters had, he thought, given him the tools he would later need to win the confidence of the 'simple people of another race in another part of the world' in order 'to use them for my scientific studies'.

Despite this sometimes rather patronising, utilitarian attitude to the groups that permitted him to observe them, Westermarck clearly loved meeting the people with whom he came into contact, delighting in showing children his collapsible opera hat and always making sure he had a plentiful supply of table fireworks with which to impress local grandees – the most successful being the 'Pharaoh's serpents'. Shy but inquisitive women plied him with questions: 'Does it rain, too, in the Christians' land?' 'Is it true that Christians live on pig's milk?' He took pleasure in telling them about the women of his country, where men only had one wife and women '"were the first to begin the meals and pick out the tidbits, whilst the men have to put up with what is left." "Oh, oh," they sighed.' But as devout Muslims, they were deeply shocked to learn that a wife would accompany her husband out of the house, displayed to the world on his arm: 'Probably it was the most indecent thing I had ever said in my life.' When old women tried to arrange marriages for him, regretfully – disingenuously – the confirmed bachelor replied that he had a dozen wives at home.

While he remained disinclined to marry, as a social institution marriage held a continuing fascination for him. *Marriage Ceremonies of Morocco* was published in 1914, building on his previous work on human relationships. As part of his research he spent a night hidden in a hole in the floor of a Berber house in order to gain 'an insight into the more intimate family life of the Berber'. Twelve years later, *Ritual and Belief in Morocco* was published in two volumes, a reflection of his interest in custom, folklore and popular faith, his free-thinking approach to organised religion and morality, and a work still considered valuable nearly a century later.

One unique aspect of Westermarck's career was his dedication to two disciplines, philosophy as well as anthropology, and to three countries, Morocco and England as well as his homeland, Finland. Growing up in a nation occupied by the Russian empire, his first thought when war broke out in 1914, even though his adopted country was a Russian ally, was 'the impassioned hope that Russia would be utterly defeated'. The sense of belonging to an occupied people must have been one source of the outsider gaze that he possessed and that unites these early fieldworkers.

Westermarck's philosophy informed his anthropology and his anthropology informed his philosophy. He wondered whether there were objective moral facts, or just facts about morality; when moral philosophy didn't offer him satisfactory answers, he turned to moral behaviour for questions like the extent to which criminals should be treated with kindness. His conclusions rejected on one hand both nihilism and a universalism that held moral rules to be objective truths, and on the other cultural relativism as we would understand it today, in which right and wrong can be seen as a matter of opinion or cultural conditioning. Values, he suggested, are the creation of an individual's acts and choices: 'Far above the vulgar idea that the right is a settled something to which everybody has to adjust his opinions, rises the conviction that it has its existence in each individual mind, proclaiming its own right to exist, and, if need be, venturing to make a stand against the whole world.'

The anthropological idea for which Westermarck is best remembered today remains his explanation of the phenomenon of reverse sexual imprinting, known as the Westermarck effect, which he outlined in his first book about marriage and the family, before he ever set foot in the field. He observed that children who have been brought up together in a family group, even if they are not related, tend very strongly not to fall in love, mate with or marry one another as adults (during his career, he also noted this phenomenon in primates). To put it another way, they are romantically or sexually desensitised to one another.

His thesis, that mating within close family groups has always been the exception rather than the rule, challenged contemporary theories of the development of social structure. Lewis Henry Morgan, an American lawyer, scholar and social theorist working in the mid to late nineteenth century, had proposed that societies evolve just as species do (roughly according to the theories of Charles Darwin and Alfred Wallace), progressing from savagery through barbarism before arriving, at last, at civilisation. Morgan's work identified kinship as the central topic of anthropology and suggested that the gradual development of the family was the impetus for the development of civilisation. He speculated that during the first stage, which he called savagery, humans had lived in a state of 'promiscuous intercourse'; the next stage was a 'communal family' arrangement, potentially matriarchal, in which it was speculated that brothers and sisters would have married, before moving on to a 'barbarian family' stage, various patriarchal (therefore better) systems including polygamy, the final step before 'civilised' human intercourse, a peculiarly Victorian conception of love, definitively patriarchal and combining emotional attachment with property while prioritising sexual exclusivity and primogeniture.

If Westermarck's theory was correct, it meant that there had never been a time of either promiscuous intercourse or communal family: from the start, humans had lived in small family groups and, on the whole, mated outside them. Today we understand the genetic imperative for this and zoologists have observed the same behaviour in our closest mammalian relatives. Kinship was thus integral to human nature and not (like culture) a social construct. Westermarck's theory was also a welcome riposte to the theories of Sigmund Freud, doing his own fieldwork in Vienna at this time, deep in the caverns of his patients' – and his own – imaginations. Freud was fascinated by unconscious sexual desires, which he believed were rooted in early childhood and centred on close family, but Westermarck's theory seemed to demonstrate that there was no anthropological basis for these ideas.

Debate around Westermarck's thesis continued to simmer for decades. One of his heroes, Sir James Frazer, observed in 1910 of the incest taboo that 'we may safely assume that crimes forbidden by law are crimes which many men have a natural

Edvard Westermarck (shown here in Tangier in 1920) adopted a dashing crimson fez during his decades of living and working in Morocco.

propensity to commit'. Westermarck responded in a 1936 lecture: 'Do the severe laws against bestiality prove that a large number of men have a strong natural propensity to copulate with animals? ... Customs and laws express the general feelings of the community and punish acts that shock them. But they do not tell us whether the inclination to commit the forbidden act is felt by many or by few.'

To the modern observer of Westermarck's life, it is hard to avoid speculating that acts forbidden in his lifetime by law as well as convention were something to which he was drawn. He never married and, even now, it is almost impossible to find any evidence of a personal life, despite his volume of memoirs; he kept his life neatly compartmentalised. At times he was the progressive social studies professor in London, at others the fez-wearing traveller in a distant land, and at still others the philosophy don and first rector (between 1918 and 1921) of the newly founded Swedish-language university Åbo Akademi in Turku, Finland. Being a bachelor allowed him both to devote himself to his work and to move easily between worlds; early success gave him the luxury of following his own scholarly instincts with no supervisor to report to and no family to support.

The central relationship of his life, apart from with his parents and sister, to whom he was always close, seems to have been with Sîdi Abdsslam, 'my shereef', his devoted Moroccan guide and companion over decades of fieldwork and exploration. He mentioned him frequently in his memoirs – he even took

Abdsslam with him on holiday to Italy – but barely described him and made no comment on their friendship. The motto by which he declared he lived, *Bene vixit qui bene latuit,** indicates how he prized his privacy.

What he did do was thoroughly question received Western morality and write with studied neutrality about cross-cultural homosexuality in his landmark *The Origin and Development of the Moral Ideas* (1906), in which he was one of the first scholars to use the phrase 'homosexual love' in its modern sense. Homosexuality, he wrote, 'is frequently met with among the lower animals. It probably occurs, at least periodically, among every race of mankind. And among some peoples it has assumed such proportions as to form a true national habit.' It occurs sometimes due to 'instinctive preference, sometimes to external conditions … Even between inversion and normal sexuality there seem to be all shades of variation.'

He noted love between comrades-in-arms, classical pederasty, and 'men who behave like women [and] … women who behave like men'; changes of sex, he explained, had been recorded in Africa, Asia and the Americas, usually accompanied by future shamanship. In Morocco, as in ancient Greece, it seemed, he observed, 'that the ignorance and dullness of Muhammedan women, which is a result of their total lack of education and secluded life, is a cause of homosexual practises; Moors are sometimes heard to defend pederasty on the plea that the company of boys, who always have

---

* Ovid: He lives well who lives out of sight.

news to tell, is so much more entertaining than the company of women'.

It was no accident that Westermarck (like his friend the sexologist Havelock Ellis, whose pioneering series of medical textbooks, *Studies in the Psychology of Sex*, came out in English in several volumes from 1897) at once denied the idea of universal primitive promiscuity while challenging the morality accepted in his lifetime. If humankind was progressing from promiscuity towards a civilised model of men and women two-by-two in neat, missionary-position pairs, as Morgan's ideas suggested, then inverts – as those who deviated from this ideal were known – were little more than savages, their primitive urges unrestrained. Instead Westermarck argued eloquently that from anthropological and historical angles homosexuality was neither especially unusual nor especially abnormal.

Work was his dominant passion. Part of the joy of sociology (of which, in Europe, anthropology was then a branch) was its new-ness as a discipline. 'Anyone who takes up the study of sociology must not expect to come to an exhibition, where every article may be had ready and finished. On the contrary, he will find that he has entered a workshop, where everything is in the making – and he will have to take part in the work.' He acknowledged the difficulties in examining social phenomena, their hidden causes, the impos-sibility of objective interpretation: a sociologist will have to deal in hypotheses, rather than truths, but 'hypotheses are legitimate in every science, and ... exactness is, after all, a question of degree'.

Just as Westermarck was ahead of his time in his choice of subject – he became the first professor of sociology in England, along with Leonard Trelawny Hobhouse, at the University of London in 1907 – he was progressive in the way he conducted his teaching. Finding the model in which a professor laid down his subject like an oracle for his students to absorb limited and old fashioned, he encouraged instead a collaboration, wherein 'both parties are engaged in a common search for truth, and it becomes evident how little we really know'.

Seminars, with students leading the discussion and the teacher questioning them, suited his more egalitarian style. He offered in his memoirs a typical list of a term's worth of seminars, which demonstrated an impressive breadth of subject and student: Miss Freire-Marreco* on Pueblo Indians in New Mexico; Major Tremearne on the Hausa language of Nigeria; Mr Wheeler on Solomon Islanders; an Indian physician, army officer and sociologist, Major Bilgrami, on East–West relations in India; Dr Oppenheimer on the criminal responsibility of lunatics; and finally, Mr Anstey discussed sacrifice – of what nature Westermarck did not specify.

The impression given by his affectionate account of the atmosphere at his department, with its mixed student body – in age, gender and nationality – meeting at restaurants after seminars, smoking together over tea (even the ladies!), and pushing back

---

* Barbara Freire-Marreco was one of the first two students at Oxford to receive a diploma in anthropology (hers with a distinction), the first anthropologist from Oxford to embark on fieldwork and the first woman to lecture in anthropology there. See Larson, *Undreamed Shores*.

the tables to dance in the dining hall, is one of conviviality and mutual respect. In his praise of his colleagues' gaiety, frankness, humour and cordiality you can get a glimpse of the man himself, passionate about his subjects and modest about his achievements. 'How heartily he can laugh, and how he loves a glass of good wine in good friends' company,' he wrote of Alexander Shand, vice-president of the Aristotelian Society, and one of the first friends he made when he arrived in London; it's not hard to imagine that description suiting Westermarck, too.

But despite his intellect, scholarship and teaching, Westermarck's legacy, even during his lifetime, was never secure. His decades-long career may have inspired the concerns and techniques of the students who followed him – one of whom, Bronislaw Malinowski, would utterly transform English anthropology – but it also reflected the preoccupations and methods of the great scholars who preceded him. Despite his refutation of the ideas of Lewis Henry Morgan, he was associated too closely with the old-fashioned classical social evolutionists for the younger generation of scholars, who were starting to question the idea that Western civilisation was the apotheosis of human endeavour. For them, classifying races or cultures as civilised or primitive was arbitrary, speculative and ethnocentric.

Like William Rivers, Westermarck recommended, for example, that colonial officials should study social anthropology to learn more about the people they governed across the world. This was a progressive view for its time, but it contained within it the assumption that Western nations had the right to govern 'savage' races and were helping to speed up their path to 'civilisation'. 'Ignorance has always been the main cause of the troubles which have followed upon the contact between different races,' he said. 'I am convinced that in our dealings with non-European races some sociological knowledge, well-applied, would generally be a more satisfactory weapon than gunpowder. It would be more humane – and cheaper too.'

One might argue that the fact that one race had gunpowder and the other didn't had more to do with those troubles than the ignorance that was shared by both but, on the whole, Westermarck was respectful of the cultures he observed. In his memoirs, he described with admiration a Native American leader, Ojijatekha Brant-Sero, heir to the chief of the Mohicans, attending a 1904 meeting of the Sociological Society. Brant-Sero emphasised the mistakes European ethnologists so often made in their investigations into the so-called 'savage' races and recommended they use 'intelligent "savages"' to aid their research. For his part, Westermarck called his Moroccan informants his 'teachers' and wrote in his memoirs that 'their patience [with him] was beyond all praise'.

Like William Rivers with the Toda and Robert Lowie with the Plains Indians, Westermarck noted with amusement the tendency of every group he encountered to view itself as superior to others. 'According to Eskimo beliefs, the first man, though made by the Great Being, was a failure, and was consequently cast aside and called *kob-lu-na*, which means "white man", but a second attempt of the Great Being resulted in the formation of a perfect man, and he was called *in-nu*, the name which the Eskimos give to themselves. Australian natives, on being asked to work, have often replied, "White fellow works, not black fellow; black fellow gentleman." When anything foolish is done the Chippewa Indians use an expression which means, "As stupid as a white man." The Japanese imitate our inventions and utilise our knowledge, but have no admiration for our civilisation as a whole or our views of life. Muhammedans envy us our weapons of destruction, but in their hearts they regard us as their inferiors. When we carefully scrutinise what other people think of us, we come to the somewhat disappointing but not altogether unwholesome conclusion, that the belief in the extreme superiority of our Western civilisation really only exists in the Western mind itself.'

Perhaps even more damning than his devotion to the idea of Western civilisation, in his successors' eyes, were Westermarck's associated 'comparative' methods. He sought out universal answers to the questions of morality and behaviour by which he was so fascinated and, in the tradition of the great nineteenth-century polymaths, was unafraid to use biological, psychological, philosophical and historical theories to complement his anthropological insights. Critics accused him of imposing his conclusions on his material rather than putting the research first; in an era that increasingly prized single studies, the new functionalist approach, he was labelled a generalist and, still worse but entirely falsely, an armchair practitioner. While acknowledging that caution must be exercised, he responded with a prescient warning that the comparative method he favoured could save ethnology 'from falling into a great number of fragments without any inherent connection, or from becoming a soulless study of the migration of phenomena without an attempt to find their real origin'.

It might be too that Westermarck's legacy faded because he was innately modest, uninterested in being lionised by his students, whom he saw as colleagues rather than disciples. (This could not be said of Boas, Radcliffe-Brown, Malinowski or Lévi-Strauss.) He was also determined to avoid being pigeonholed – he liked to quote Montaigne, whom he called 'a pioneer as a student of savage customs': 'to be wedded to no theory is a warning that every ethnographer should have at heart'. Never believe anything until you have seen it for yourself.

The fieldwork fetishists within the discipline as it evolved may not have fully recognised Westermarck's contribution to it. But although his broader works were comparative, his Moroccan ethnography was immersive and scrupulously observed and recorded. It contributed to his anthropological and philosophical thinking on marriage and the family, on religion and morals, on pagan survivals in modern ritual, on migration and magic. While no single period of fieldwork defined his career, as it did some of the men and women in this book, almost uniquely he developed a lifetime respect for and attachment to the country in which he worked for so many years, travelling there for friendship, relaxation and pleasure as well as for work.

Five of Westermarck's Moroccan travelling companions, including his friend the Shereef (second from right), in a field of wild flowers.

Morocco, in its foreignness and warmth, its lack of modern amenities and the luxury of its beauty and space, seems to have answered an undefined yearning in Westermarck's soul. Beneath the dapper exterior of a benevolent professor, he can be seen as a secret rebel, free-thinking, unconventional, wild at heart. Morocco liberated him from something undefined, perhaps something even he preferred not to acknowledge. Climbing in the High Atlas above Marrakesh with his friend, he wrote, 'the shereef [*sherif* or lord in Arabic: his name for Sîdi Abdsslam] complains of backache, but that is no good. My lungs are filled with air, my limbs twitch with delight. Here I am happy, here I am free.'

CHAPTER FOUR

The Magi

Daisy Bates and Alfred Radcliffe-Brown
in Western Australia, 1910–1912

———•———

In the southern hemisphere winter (European summer) of 1914, a small army of British scientists descended on Australia for the eighty-fourth meeting of the British Association for the Advancement of Science (now called the British Association). Several celebrated anthropologists spoke on the final day of the proceedings in Sydney, as recorded in an article in the 22 October 1914 edition of *Nature*: after Professor Haddon spoke on the importance for the colonial administrator of the study of anthropology and Professor Rivers discussed the influence of gerontocracy on marriage among Australian Aboriginals, 'Mr A. R. Brown followed with an account of the varieties of totemism in Australia, his classification covering several new types recently discovered by himself in Northern Territory, or by Mrs Bates in the Eucla district.'

Mrs Bates – Daisy Bates – gave a rather more loaded account of the morning's lectures. She had spent over a year in the

Australian bush with Alfred Radcliffe-Brown* and she believed he had stolen her research; travelling by camel for two weeks across the Australian southern desert from her remote campsite, she'd come to the lecture to expose him. His paper, she told a friend indignantly, 'was all an extract from my MS. At the close of Brown's paper, Sir Everard im Thurn [president of the Association] asked me if I would like to add to Mr Brown's paper. I said that Mr Brown had given my notes so nicely there was no need to add to them. The meeting "grimaced audibly" and I don't think Brown will ever forget the incident.' But any jubilance at this tart retort was short-lived. Bates's one-time colleague Radcliffe-Brown, then a promising but untenured anthropologist, would go on to enjoy a glittering career as a scholar, while Bates would give up any hope of academic recognition and return to her life's work, defending and protecting her beloved 'natives'.

Four years earlier, Bates had been overjoyed to hear that the University of Cambridge was funding Radcliffe-Brown's expedition to Western Australia. In February 1910, she wrote to a fellow amateur ethnologist, Reverend John Mathew, of her excitement about accompanying Radcliffe-Brown and his assistant to meet the Aboriginal people with whom she had been working closely for a decade. 'As Professor [*sic*] Brown is a young man under thirty, I have placed him in the Paljeri Division where he

---

* Radcliffe-Brown was born Alfred Reginald Brown in 1881 and changed his name by deed poll to Alfred Radcliffe-Brown in 1926. Radcliffe was his mother's maiden name. Although he was still called Brown during the events described, I will refer to him as Radcliffe-Brown because he is best known by that name today.

becomes my son. (I am Boorong [tribe] throughout the north so one difficulty is removed as a mother can accompany her son anywhere.)' She believed she would be Radcliffe-Brown's guide and mentor but he had other ideas.

While Bates was writing to Mathew, Alfred Radcliffe-Brown and his assistant, 'Peter' Grant Watson, were making their arrangements for the sea voyage to Australia. A scholarship boy, Radcliffe-Brown had begun studying natural sciences at Trinity College, Cambridge, in 1901 but, under the influence of Alfred Haddon and William Rivers, he'd moved his focus to anthropology. Handsome, intense and charismatic, he had earned himself the nickname 'Anarchy' Brown due to his undergraduate flirtation with the ideas of Peter Kropotkin, who, he observed, had 'pointed out how important for any attempt to improve society was a scientific understanding of it'. Radcliffe-Brown's radicalism was part of a contrived persona which included, as costume, a flowing cape, high hat and gold-topped cane.

He had spent the years between 1906 and 1908 researching the people and culture of Great Andaman* in the Indian Ocean, a tropical paradise where few travellers over the ages have dared venture. Even Marco Polo had recorded that the Andaman islanders ate strangers to their shores. Radcliffe-Brown, though, had apparently persuaded his hosts to see him as 'a being from

---

* He spoke Great Andaman, but without an interpreter couldn't learn the language of Little Andaman, the more remote of the two islands. Great Andaman islanders had iron, probably gleaned from shipwrecks, as well as syphilis and measles, which indicated some contact with sailors.

Alfred Radcliffe-Brown, radical young social scientist, c. 1910. A colleague described him as 'impenetrably wrapped in his own conceit'.

another world: no white man, but ... a black man resurrected from the happy hunting grounds of the dead, sent back to them to be their protector and their master-spirit'. Grant Watson (who got his information direct) said Radcliffe-Brown had lived there 'as a primitive autocrat, exercising a beneficent but completely authoritarian sway over the simple Andamanese'.

Despite his rather Aleister Crowley-esque image (the occultist was at Trinity a few years ahead of him) and his acceptance of the need to be viewed by his subjects as a deity, Radcliffe-Brown intended his work to be pure and rigorous. 'It is clear that such studies [fieldwork] need to be based on a scientific and carefully elaborated method,' he wrote in his 1922 book on the Andamans, echoing his mentor William Rivers, to whom (along with Haddon) the book was dedicated. 'Unfortunately ethnologists are not yet agreed as to the methods of their science.' Claiming to have introduced the ideas of the influential French sociologist Émile Durkheim to English academic thought, Radcliffe-Brown was determined to change all that.

The future novelist E. L. Grant Watson, known as Peter, had just received his degree in zoology at Trinity in 1909 when Radcliffe-Brown asked him to join him on his expedition to Australia. Grant Watson was to fund himself and, when Radcliffe-Brown delayed his departure at the last minute, he travelled alone, accompanied instead by Halley's Comet, which blazed in the night sky throughout his long voyage south. Neither Radcliffe-Brown's absence nor the fact that he was paying his own expenses dimmed Grant Watson's admiration for his mentor, whom he described as brilliant but maverick, wild and iconoclastic. He didn't always live up to his own high standards – he could be overbearing, rude and ruthless – but Grant Watson also found him stimulating, beautiful, spellbinding. 'One must cultivate style. He dressed like a Paris *savant*, faultlessly. He aspired to be conscious at every gesture ... and I have never known any man read poetry better.'

Others thought him a charlatan and even Grant Watson had to confess that women especially considered him 'no gentleman'. Three American anthropologists who came into contact with him in the 1920s and 1930s confirmed this. Hortense Powdermaker commented on his need to be worshipped; Margaret Mead said he took 'any disagreement, tacit or uttered … as a slap in the face'; Ruth Benedict, perhaps thinking of his cloak, described him as 'impenetrably wrapped in his own conceit'. When Robert Lowie, one of Franz Boas's disciples, published an unflattering portrait of Radcliffe-Brown in 1937, it prompted a sulky, intemperate letter of complaint.

In 1910, though, Radcliffe-Brown was just an eccentric but promising young anthropologist, much less well known than Daisy Bates. In her own way, she too was a self-invented, self-dramatising figure. Born Margaret Dwyer or O'Dwyer in Tipperary in 1859, her accounts of her life were both rather more romantic, and rather more respectable, than the limited surviving documents suggest. Her father, she always insisted, had been an impoverished gentleman, though he had a fondness for the bottle; her mother died young and she was brought up by a grandmother and an Irish nanny – and in various orphanages, though she seldom mentioned that.

What she took from her childhood was a love of simple, uneducated people who believed in spirits and the magic of landscape – she used to say that the humble Irish men and women with whom she grew up had 'little to separate them from the Australian aborigine' – and a passion for the British Empire. She was a headstrong girl, mercurial and proud, with a sharp but untrained intellect, who loved riding to hounds and dancing reels. In 1882, fleeing rumours of an unsuitable attachment, she went to Australia to stay with the Bishop of North Queensland, a friend of the family, and look for work as a governess.

Opportunity and adventure opened up in front of her: Australia's vast spaces were big enough to contain her vitality. Everyone was new there; everyone seemed to feel their potential expand under those deep blue skies. The white settlers, at any rate. 'Blackfellas', the kindest name by which Australia's ancient indigenous people were known, were seen by the new arrivals as 'more like spiders than anything human': a pest to be eradicated, swept into distant corners, while the important business of exploration, expansion and exploitation got under way. From the 1860s onwards, Australian state governments sought to control and regulate the lives of Aboriginal people, specifying where they could live and work and reserving the right to take their children – particularly 'half-caste' children, as they were known – from their parents for their own 'protection'; these powers would not be repealed for a century.

In 1884, at the age of twenty-five, Daisy married a handsome young drover, or stockman, Harry 'Breaker' Morant.* Like her,

---

* Breaker Morant (1864–1902) was a drover, horseman and bush poet, a friend of Banjo Paterson. He was executed by firing squad after being court-martialled for killing prisoners of war in South Africa during the Boer War. The circumstances were muddy and he and his companion, Peter Handcock, became folk heroes, early symbols of Australian defiance of British rule. Both refused to wear the blindfolds they were offered at their execution and Morant's last words were reported to be 'Shoot straight, you bastards! Don't make a mess of it.'

he was new to Australia and happy to polish up his account of his past, claiming to be the illegitimate son of an English admiral. He also claimed to be twenty when he was in fact nineteen – because of which their marriage, which lasted a matter of weeks, was never legally valid. Daisy left him because he didn't pay for the wedding and stole some pigs and a saddle but the marriage was never dissolved.

Her next union was with Jack Bates, another bushman, across the country in New South Wales less than a year later; four months after that, in Sydney, and still without dissolving any previous attachments, she secretly married Ernest Bagehole, a seaman from a wealthy London family whom she had met on her voyage to Australia. Either Bagehole or Bates was the father of her only child, Arnold Hamilton Bates, born in Bathurst, New South Wales, in August 1886, though Bagehole soon faded out of her life. For the next seven years, rather than staying with her Bates mother-in-law when Jack was away droving, she took Arnold with her while she worked as a governess in the homes of far-flung farmers.

In 1894, Daisy Bates left eight-year-old Arnold at boarding school in New South Wales, spending holidays with his grandmother, and returned to Britain to find work as a journalist. She was employed in London by the editor W. T. Stead as an occasional daily at the *Review of Reviews*, a monthly journal intended to synthesise the best journalism across the British Empire. Stead was one of the first editors to hire female journalists and his campaigns against poverty and child prostitution were reflected in changes of public opinion and the law; he also invented the 24-point headline and, when he profiled General Gordon in 1884, introduced the interview to modern journalism.

A letter to *The Times* about the conflict between Western Australian settlers and the indigenous people whose lands they were seizing sparked Bates's return. The settlers wondered who were the masters, '"us" or the natives', while the Aborigines protested that 'the smell of the white man is killing us'. Bates wrote to *The Times* offering to investigate the situation for them and they accepted, paying her passage back to Australia the same year, though it seems no resulting article ever appeared. By this time, around 1899, Jack Bates had acquired a farm in Western Australia, Ethel Creek, where he and Arnold were living; when Daisy returned, the three of them bought 770 head of young cattle in Broome and, over six months, on horseback and living in bush camps, drove them 4,000 kilometres through the wilderness to Perth to sell them.

But Bates didn't want to be a farmer's wife. When Jack sold Ethel Creek and Arnold began an apprenticeship, Daisy set off from Perth into the vast and empty north-west corner of Australia, studded with majestic red sandstone outcrops, to study the Aboriginal tribes there. She believed they were a dying race, doomed to extinction at the hands of progress – progress in the form of aggressive European settlers. A conscious witness to a vanishing world, she began recording their languages, beliefs, customs and traditions: 'To live among them, to see them amidst their own bush surroundings and to note their everyday comings and goings is to experience an ever delightful feeling that you are watching the doings and listening to the conversation of early mankind.' When she could, she wrote an article, always arguing their cause, for one of the Australian or British newspapers. Unusually for a self-taught ethnologist of the time, she had no interest whatsoever in converting her companions to Christianity. It didn't take her long to realise she had found a special kind of happiness among these new friends. With her gift for languages, she picked up their multiplicity of dialects easily; she loved their warm hospitability, the joy they took in singing and dancing, their gentleness and their merriness. The immensity of their landscape moved her profoundly, as it did them.

In 1904, Malcolm Fraser, Western Australia's registrar general, gave Bates a job recording the customs and dialects of the indigenous people of Western Australia, for which she was to be paid eight shillings a day. These were the happiest years of her life, during which, although she lived austerely in the bush, she represented the government as an ethnologist, was considered the journalistic authority on Aboriginal people and, when she visited Perth, was invited to Government House receptions for the great and good; despite her disdain for convention, she was always susceptible to the gilt trappings of empire. In 1905, she presented a paper entitled 'West Australian Aborigines – Marriage Laws and Customs' to the Royal Geographical Society of Australasia.

Daisy Bates at the centre of a group of Aboriginal women. She always dressed immaculately in the costume of a Victorian gentlewoman of her youth.

In 1907, she was elected a fellow of the Royal Anthropological Society of Australasia. Between June and December 1908, she covered over 5,000 miles of bush, walking and riding as well as by train, visiting 70 towns and recording 34 'new' dialects. 'The epoch-making work Mrs Bates has done among the aborigines, the interest she has created in these quaint, lovable and primitive people to whom, when all is said and done, Australia really belongs, entitles her to rank among the women who have accomplished great things in this age,' reported the *Western Mail* in 1910.

Out in the bush, Bates lived in a large canvas tent, fourteen feet across, pitched in a windbreak of scrubby trees. She slept on a stretcher on one side of the tent, her kerosene supply and medicine chest at its foot, with her desk – a folding table – set up across the open back flaps and her portmanteau, containing her wardrobe, beneath it. On the other side of the tent from the bed were her holdall and tucker box. Her bush kitchen had a little fire pit and upturned cases for a table and stool, which she liked to use as an outdoor study; her tea towels and washing-up bowl hung from nails hammered into a tree.

She slept lightly and rose before dawn, prodding the embers of her fire into a blaze on which to set the billy, or kettle. Her main meal was breakfast – a boiled egg, bacon and bread and butter, always with tea – as she contemplated 'the slow sweet living ways of old time' and the beauty of the wilderness around her: the rising sun casting intricate shadows in the gum leaves, spider webs iridescent with dew, the vivid yellow of the wattle blossom.

Bates's hosts seem to have viewed her simply as foreign, rather than male or female. A woman in Aboriginal culture, in her words, could be bought and sold by her husband, even lent to other men; she built the shelters, fetched the firewood, carried 'the water and the baby and the spare spears and the boomerangs of her man. The native woman is today what she was 10,000 years ago – obedience personified. A touch on the shoulder with a spear and she rises at once, whatever her occupation, and goes where she is told.' She admired their endurance but relished her own independence.

This was the period in which the persona of Daisy Bates crystallised into the image we recognise today: a slight, upright, strangely ageless woman in a stiffly boned monochrome dress, often with a tightly rolled umbrella in the style of Mary Poppins, standing in front of her canvas tent on the red earth with a circle of almost naked brown-skinned men and women around her. She clung resolutely to her Victorian standards (old-fashioned even in 1910) but liked to think of herself as having white skin with a 'black man's mind'. Like Radcliffe-Brown in the Andamans, who thought of himself as 'a being from another world ... [the islanders'] protector and master-spirit', Bates, too, cultivated a quasi-magical aura, persuading the Aborigines she had come from the Dreamtime – which, since ghosts in indigenous mythology are pale, was a reasonable assumption. Inventing an identity as a shaman, she was permitted to witness the most intimate of tribal ceremonies; her adoption into the rain totem clan was an almost unheard-of honour for a European.

The simple medicines she dispensed to the sick and wounded enhanced this supernatural mystique. At some point she acquired a *nowinning*, or magic stick, about four inches long and carved with the figure of a woman. This identified her as a *mobburn*, or magic doctor, who could cure or kill anyone; Bates said she had received it from a dying *mobburn* who had chosen her as its recipient. The tribal people believed there was fire in this stick and, when she was away from the tent, they told her light shone all night from it. While she possessed this *nowinning*, no one could approach her without her permission but she could go anywhere and ask anything.

As Grant Watson put it, when he and Radcliffe-Brown arrived in Australia, Daisy Bates, then fifty-one, was known for knowing 'more about Australian aborigines than anyone else alive'. She may not have been university-trained but she was an 'enthusiast, who has given all her love and sympathy to these outcasts from their own lands' and, in so doing, earned their trust and respect. For her part, Bates had her own reasons for wanting to help Radcliffe-Brown with his research. She had been amassing observations on the tribes of Western Australia and their dialects since 1900 and been told that if she joined the expedition, not only would her expenses be covered but her work would be included in the resulting book. Radcliffe-Brown seems to have been unaware of these arrangements and, gallingly, when an admirer of Bates's (a sheep farmer named Sam MacKay) donated £1,000 to their expedition, he claimed the credit.

Grant Watson's description, written as a memoir in the 1940s, is the fullest surviving account of their fifteen months together in the Upper Murchison and on Bernier Island. It was, he commented, 'unfortunate' that Bates, for all her knowledge and courage, imagined she could lead their party. Prepared to be deferential in what they would all have called the civilised world – as women were of course expected to be – in the bush she had all the experience: she was astonished to find that Radcliffe-Brown, twenty years her junior, considered her more an encumbrance than anything else. 'She was made for his exasperation, as he for hers,' observed Grant Watson. Furiously, and with devastating accuracy, Radcliffe-Brown deployed his 'weapons of silence and aloofness' and all Bates could do was shake out 'the feathers of her soul' and confide to Grant Watson that his colleague was 'a most extraordinary man'.

Radcliffe-Brown's hostility wasn't because she was a woman, Grant Watson insisted; it was because her encyclopedic knowledge of Aboriginal culture 'was not in a condition that Brown considered easily available for the ends of science'. (One might argue that as a woman born in the 1850s, Bates had had very little chance to learn what Radcliffe-Brown considered correct scientific method, even were such a thing established in anthropology.) Furthermore, he considered her overly emotionally involved: her stated aim was to help Aboriginal people, while to Radcliffe-Brown, scientifically detached, they were merely the objects of his research. By the time they'd realised harmony would be impossible, they were marooned together a thousand kilometres from Perth, with Radcliffe-Brown muttering ominously 'that if women claimed the privilege of men, they would be treated as men in like circumstance would be treated'.

At their first camp, east of Sandstone, Radcliffe-Brown and Grant Watson settled down to investigate how Aboriginal marriage practices fitted into 'the four-class marriage system'. Before dawn one day, the morning after a night-time excursion to witness an initiation ceremony, ten or twelve white Australians galloped into the camp, 'firing off their revolvers at the native dogs, and shouting and swearing in quite a cinematograph manner. Natives were making off in all directions.' A policeman explained they were looking for some 'tribal murderers' who were said have been at the ceremony the previous night. Icily Radcliffe-Brown explained to the posse that his work was now effectively ruined, since the 'natives', upset by this outrage, would melt away into the bush, 'as only natives can'.

In the end, the police carried away a 'white haired and respectable old gentleman' (entirely innocent, added Grant Watson, as well as completely naked), and it transpired that Radcliffe-Brown was hiding the miscreants in his tent. Bates was pleased that he had 'sheltered her beloved natives from the injustice of the law' but less so to discover that he was determined to set off for another location, believing that the people they had been investigating would refuse to return. She insisted they would come back and, anyway, she still had work to do there; Radcliffe-Brown declared she could stay on alone.

Leaving an incredulous Bates behind, Radcliffe-Brown and Grant Watson moved on to Bernier Island, high up on the north-west coast, where there was 'a lock-hospital for venereally infected natives'. Over the past century, white men had brought syphilis and other dreadful diseases to the defenceless Aboriginal people of Australia and, to prevent white men being reinfected, the native sick were captured, brought through the bush in chains and imprisoned on uninhabited islands. 'These journeys, from start to finish, often took weeks; often the patients died by the way,' wrote Grant Watson. 'Flies buzzed above their suppurating sores; their chains were never removed. Men, women and children were mingled indiscriminately, and it would be a wonder if not all the survivors were thoroughly infected with all possible varieties of venereal ailment by the time of their arrival at their respective destinations.' Despite the desperation of its inhabitants, Radcliffe-Brown found Bernier Island an excellent place for anthropological research because the people came from so many tribes, allowing different stories and accounts to be collected easily and compared to one another on the spot. By all accounts he was unmoved by the stench of death.

The two Englishmen were enchanted by the island. 'The blue, sparkling tropic sea, the stiffening breeze, the sense of adventure, the plunge to the unknown': they watched the fins of basking sharks breaking through the water and vast turtles surfacing through bubbles and then plunging into the depths again with deep gurgles. Although they were plagued by ants – which despite hot ashes regularly being poured into their nests still attacked their toenails while they slept – as well as ticks and scorpions, the flies that made life inland a misery were mercifully blown clear and swimming was a daily pleasure. Unlike their subjects, Radcliffe-Brown and Grant Watson spent their first weeks on the island in a 'rhythm of good health and hot sunshine', until the day on which a cross and bothered Daisy Bates, clad despite the tropical weather in her usual pin-neat black suit and veil-swathed straw boater, stepped ashore and became even more angry at finding them so peacefully established in their work. She stayed for eight months, translating for Radcliffe-Brown, unable (it seems) to quite believe that he would use her knowledge and skills without giving her credit or sharing her concern for the people she knew so well.

Grant Watson left first, to continue his voyage round the world. The feud between Radcliffe-Brown and Bates aside, he had been profoundly influenced by his time in Australia, daily witness to what he called 'the power of magic', which would recur in his writing for the rest of his life: 'I have lived for a considerable number of months in the world of magic-sticks and stones, of totem animals, and if I have not actually met and conversed with the Alcheringa, animal ancestors, I have become convinced of their existence in the same way in which Dr Jung is convinced of the existence of the archetypes ... That I have seen stark young women streaming with the blood of a yet living turtle which they were laboriously dismembering with a stone knife, is a picture every bit as significant as the intellectual tea-parties I have attended at Lady Ottaline Morall's [*sic*].'

Radcliffe-Brown and Bates laboured on in separate dignity until early 1912. Radcliffe-Brown was still in Australia in 1914, when, stranded by the outbreak of war, he taught at Sydney Grammar School before taking up a teaching post in Tonga between 1916 and 1919. Bates, the first woman to have applied to become the Northern Territory's official paid Protector of Aborigines, was disappointed instead to be given the unpaid role of Honorary Protector of Aborigines at Eucla, the southern- and eastern-most region of Western Australia, where she remained for sixteen months, rebuilding her confidence and motivation.

Although they had separated bitterly, she must have been in contact with Radcliffe-Brown from there because his convention speech in Sydney in the winter of 1914 drew on her Eucla research.

She had given the second copy of her ethnographic manuscript to Radcliffe-Brown. The first copy had gone to the Scots poet and folklorist Andrew Lang, author of a series of cloying 'fairy books' that retold folk tales from around the known world, including, in the 1904 *Brown Fairy Book*, Australia. Before Radcliffe-Brown's expedition, Lang had promised to ask William Rivers for help on Bates's behalf but, afterwards, Radcliffe-Brown outraged Bates by telling her Lang had called for a 'red pencil' for her 'long and wandering work'. Lang died in July 1912, before he had the chance to wield his red pencil. Radcliffe-Brown didn't even deign to get his out: the copy Bates had given him was left behind when he departed Australia, in her words, 'mutilated and useless'.

Having failed in her scientific role, Bates reaffirmed her intention of becoming the protector and guardian of her Aboriginal friends rather than their scholar: 'My sole desire is just to live among my black friends ... I have not a particle of personal ambition or self-serving beyond my desire to impress upon this native race that there is one woman who is absolutely their friend, without thought of self-advancement. When all look upon me as a friend, when men, women and child [*sic*] come to me freely and confidently ... well then here my mission lies.'

Radcliffe-Brown didn't write up his Australian research until 1931 – a brilliant lecturer, he found writing agonising; it took him

Bates in 1932 at Ooldea, the place she considered home. Note her heavy, fly-repelling veil.

more than ten years to publish his Andaman research – but he drew on it throughout his career. His broader anthropological ambition, to which his limited fieldwork (only these two early periods in the Andamans and Australia) was directed, was to explain social phenomena not in psychological or historical terms, but scientifically, according to abstract, discoverable and classificatory conditions and laws. He conceived of societies as systems of interdependent elements, functionally allied – hence that anthropological watchword, 'functionalism', though (typically) he vigorously refuted any link with the movement more closely associated with Bronislaw Malinowski (see Chapter 5). Critics called his work sparse and rigid; followers admired his empiricism and academic purity. The austerity he hoped to convey in his work strikes a stark contrast with his flamboyance, the personal mystique he quite deliberately cultivated.

Unusually among his contemporaries, Radcliffe-Brown took university posts across the world between 1914 and his death in 1955: Cape Town, Sydney, Chicago, Oxford and Alexandria, with visiting professorships and occasional teaching positions in Yenching and São Paulo as well as London, Manchester, Birmingham and Cambridge. He founded the journal *Oceania* in 1930. He was an inspirational and compelling teacher but academically confrontational, as reluctant to acknowledge the work of previous researchers as he was to admit any criticism of his own. When he left England for Australia in 1910, he was newly married, though Grant Watson didn't mention his wife; they had a daughter but were estranged by the mid 1920s and divorced in 1938. Few images of him survive and almost no trace of a personal life.

For the rest of her life, Bates moved between Adelaide and remote camps in Western Australia, still working with her Aboriginal friends, still scraping a living from journalism. She became known as *kabbarli*, a word that means grandmother but can also mean granddaughter. In 1934, she was made a Commander of the Order of the British Empire. In 1938, she published *The Passing of the Aborigines*, written with the help of Adelaide journalist Ernestine Hill, which appalled white Australians with its matter-of-fact accounts of tribal infanticide and cannibalism, prompted (according to Bates) by starvation due to the loss of traditional foods. During the war, she tried to make contact with her son, Arnold, who was living in New Zealand, but he refused to speak to her. She died aged ninety-one in 1951 and lies in the cemetery in Adelaide. It is sad to think that she was not buried at Ooldea (a remote railway siding on the Trans-Australia Railway, where Bates lived for sixteen years), as she wished, which she thought of as her home.

In her seventies, she recorded a radio interview that Julia Blackburn, author of the most evocative book about her, *Daisy Bates in the Desert* (1994), has quoted from, noting Bates's soft voice and residual Irish lilt. Evasive about her past and the details of her life, she was eager to talk about the people who had shaped it in words that reveal her self-taught method and passionate personal involvement in her work, the opposite of the rational, objective academic style of her counterpart, Radcliffe-Brown: 'They died out so quickly and I have been watching them die out. Poor women, young women, girls, men and children, I saw them turn over on their sides and die, with their bodies still young, still fine, but the breath gone from them … I gathered up their legends, everything connected with them; data, yes, that's the word. Dialects. Every bit of it is straight, true. I got it from them, I saw it done. I would sit there at three o'clock in the morning drinking from my pannikin of tea, after the whole of the proceedings – getting all their dialogues, dialects, their marriage laws and so on and so on and their legends, you know … They loved to hear me say their own words in their own way ... And I kept it up and kept it up. I never failed them, no, not for one hour of my time with them. I would go back tomorrow. That was my life. All my time among them was my life.'

CHAPTER FIVE

The Hero

Bronislaw Malinowski in
the Trobriand Islands, 1915–1917

———•———

'Imagine yourself suddenly set down surrounded by all your gear, alone on a tropical beach close to a native village, while the launch or dinghy that has brought you sails away out of sight …' This classic opening to an account of fieldwork sounds almost more like the introduction to an adventure novel, in which the hero – Bronislaw Malinowski – triumphs over loneliness and his own frailties to bring back to civilisation the wisdom he has gained from his time with primitive man. It is wisdom the savages (as he insisted on calling them) themselves could not impart; he alone, the intrepid anthropologist, can interpret it. Malinowski may have liked to joke about being Joseph Conrad, but looking back, he seems irresistibly like Indiana Jones, shadowed by students with 'I love you' inscribed in kohl on their fluttering eyelids.

Bronislaw Malinowski had arrived to study anthropology at the London School of Economics in 1910, aged twenty-six. After a peripatetic childhood based in Austrian-controlled Poland – his father, a professor, died when he was fourteen and he and his mother travelled frequently because his health was delicate – he had studied physics, mathematics and psychology at university in Krakow before receiving a doctorate in philosophy. A talented linguist, he read Latin and Greek and spoke excellent Italian, German, Spanish and French as well as English and Polish. His supervisors at LSE were Charles Seligman, who had accompanied Alfred Haddon and William Rivers to the Torres Strait twelve years earlier, and Edvard Westermarck; their anthropology department was only a couple of years old and its reading list contained just four books.*

Malinowski researched his thesis and first book, *The Family Among the Australian Aborigines* (1913) – like Westermarck's – beneath the domed ceiling of the British Library's Reading Room, at the heart, then, of the British Museum. He was also, as was his wont during this period, unhappily in love with not one but two married women. Toska (as he referred to her in his diary) worked in the British Museum, probably researching for her husband; her face was to him 'the embodiment of the feminine ideal'. But at other times, 'when looking at the Egyptian women [figures in the museum], at the sacred symbols of marriage and motherhood, a hopeless longing for Zenia sweeps over me'.

---

* E. B. Tylor's *Primitive Culture* (1871), A. H. Keane's *Ethnology* (1896), Joseph Deniker's *The Races of Man* (1900) and Alfred Haddon's *The Races of Man* (1909).

But given anthropology's developing emphasis on fieldwork, armchair research was never going to satisfy Malinowski's ambitions. The 'primitive world' beckoned. Seligman tried and failed to get funding for his star student to go to Sudan. Instead, in late June 1914, planning to continue on to New Guinea, which had been the focus of Seligman's *The Melanesians of British New Guinea* (1910), Malinowski joined the British Association for the Advancement of Science's Australian jamboree. Peopled by a great crowd of academic colleagues – of more than 300 delegates, over half were part of the paid official party, including William Rivers, Alfred Haddon and Charles Seligman representing the anthropologists – the various sea journeys south became an 'intellectual pleasure cruise'.

Advised by Seligman, Malinowski was well prepared for an extended journey. In his luggage were twenty-four crates of tinned food˙ from the Army & Navy Stores, two dozen custom-made notebooks lined with thick cream paper, two new Norfolk jackets with matching breeches and a pair of matching 'knickers', or shorts, as well as an extensive array of lamps, camp tables and chairs, a canvas bath, a camp bed, a tent (which proved too small) and a parasol from the colonial outfitter Lawn & Alder. He estimated he'd spent £146 equipping himself for a year in the field, at a time

---

* As Malinowski's biographer Michael Young notes, his supplies were Rabelaisian: lemonade crystals, tinned oysters and lobster, various kinds of chocolate and cocoa, Spanish olives, cod roes, jugged hare, tinned and dried vegetables, half-hams, French brandy, tea, six different kinds of jam and plenty of condensed milk. His medical kit contained almost five thousand tablets but he took only one toothbrush.

when the average income was about £100. Despite worries about how his health would hold up in a tropical climate, he felt himself to be sailing towards the intellectual destiny he craved.

His diary bears witness to his passion for travel for its own sake. After Naples, his ship docked in Port Said: 'Street bazaar, illuminated; wonderful smell of trees and ghastly stink of the streets; large dark houses with circular balconies. From far away Arabic music can be heard.' The heat about which he was so apprehensive grew more and more intense. His clothes were still damp with the previous day's sweat when he put them on each morning; underwear he dispensed with entirely. In Sri Lanka (then Ceylon) the untried ethnologist was a rapturous witness for the first time to 'a magic ceremony conducted by people of another culture'.

When war broke out in August, Malinowski had been in Australia only a matter of weeks. As a subject of the Austro-Hungarian Empire, he was considered by the British an 'enemy alien', too dangerous to return to Europe; he only just escaped internment. Alfred Haddon, ever helpful to promising young anthropologists, pulled some strings, introducing him to Atlee Hunt, Secretary to the Department of External Affairs of the Commonwealth Government of Australia. A progressive civil servant who believed in the practical value of anthropological research, Hunt took Malinowski under his wing, defending him from rumours of sympathising with the Germans, and provided him with a travel pass to Papua New Guinea (then a British territory) and generous grants.

By September, Malinowski, freshly supplied with medicines ('cocaine, morphine, emetics'), was waving his handkerchief from the deck in farewell to his Brisbane hosts and feeling he 'was taking leave of civilisation'. Days later he was in Port Moresby, New Guinea, en route for the small island of Mailu, where he would embark on his first extended period of fieldwork. Like Franz Boas, despite the strangers all around him, he saw himself as setting off alone into the wilderness; he was, he wrote on the opening page of a new diary, beginning 'a new epoch in my life: an expedition all my own to the tropics'.

Although these few months in and around Mailu were more survey than immersion, they served as a valuable apprenticeship for Malinowski. He was, as he wrote to Seligman, 'undergoing a practical training in your school ... I have a concrete idea of what the difficulties [of working in the field] will look like and I have lost my original diffidence.' Returning to Adelaide for the Australian autumn in March 1915, he wrote up his notes at racing speed while seducing his host's daughter, Nina Stirling. The prompt publication of his research in the *Transactions and Proceedings of the Royal Society of South Australia* (December 1915) ensured the funding Atlee Hunt had helped him obtain would continue.

In the spring and summer of 1915–16, Malinowski went for the first time to the Trobriand Islands, a string of paradisal coral atolls off the eastern coast of New Guinea. Although his later description of arriving there created the impression that he was the first European man to get to know these islanders well, this was by no means accurate. Missionaries had been active in Melanesia since the 1860s – bringing the islanders cricket as well as Christianity, and producing the first English–Melanesian dictionaries and anthropological works – and two of Malinowski's mentors, Charles Seligman and William Rivers, had studied there in the field in the previous decade. European traders dealt in pearls and exotic feathers, farmed tobacco and investigated mining opportunities.

By early 1916, Malinowski was back in Australia, living in Melbourne; he had abandoned his diary for the moment. It was at a dinner on 'Professors' Row', a street inhabited by the academics of Melbourne University, that he met Elsie Masson, the twenty-six-year-old daughter of a distinguished chemistry don. A year as an au pair in a remote cattle station in the Northern Territories had given her a passionate interest in Aboriginal people and culture and the letters and articles she had written there were published in 1915 as *An Untamed Country*. She was training as a nurse and had recently heard that her fiancé had died at Gallipoli. Even so far away from the devastation and savagery of the European war, it was a period of great intensity, especially for young people, as Elsie wrote: 'It seemed to me more than ever as if everything of one's youth was being snatched away from one and there was to be nothing left … I also … began to have a conviction that I felt very differently about the war to Mother and Father.'

Elsie described Malinowski at the time of their meeting as 'very' foreign-looking, 'a tall man with a square reddish beard, big glasses, an impenetrable expression'; grave, intellectual, with impeccable manners. When a friend suggested she go to see his collection of objects from New Guinea, 'on the score of my interest in "blacks"', she went, she said, because she was curious about him. She imagined him as the kind of man who 'had been with countesses and milkmaids in an experimental kind of way'; she also thought he looked lonely. Gradually, despite Malinowski's unofficial engagement to Nina Stirling, he and Elsie became closer. In March 1917, Bronio (as she called him) was in hospital with an infection. One of his nurses was a friend of Elsie's, who told her she'd asked him if he chased butterflies; he replied, 'No, I chase girls – it is the same thing.' Elsie was amused. Malinowski was no longer the dignified scholar she had first encountered; she found him witty, stimulating, spiritual – and clearly, although she didn't use the word, sexy.

Theirs was a union of intellects and ideals. In August, when Elsie was briefly away, Malinowski wrote his first letter to her: 'My work is absolutely soaked with your personality and whenever I plan something or write something down I find that I am addressing you.' He set off for his second expedition to the Trobriand Islands in October, beginning his diary of self-improvement again.* 'I am really not looking forward to the long and monotonous stay in the

---

* 'Day by day without exception I shall record the events of my life in chronological order. Every day an account of the previous day: a mirror of the events, a moral evaluation, location of the mainsprings of my life, a plan for the next day. The over-all plan depends above all on my state of health. At present, if I am strong enough, I must devote myself to my work, to being faithful to my fiancée [Elsie, not Nina], and to the goal of adding depth to my life as well as my work.'

Trobriands in itself, but as a means to an end,' he told Elsie at the start of his journey. Like Boas in the Arctic, even the landscape reminded him of the love he was leaving behind. 'This enormous long stretch of hilly shore makes me feel almost impatient at so much beauty wasted. Every stretch of sand, enclosed between two headlands, every little island, every promontory, unfolding its maze of mysterious coves and valleys, as we approach seems to be almost awaiting "me and you", almost an embodiment of a forgotten dream,' he wrote. 'It is like listening to music of which you could not seize the meaning yet would fascinate you.'

Once he had set up his camp at Kiriwina in the north of Boyowa Island, his days began to run into one another: an early start, vigorous Swedish 'gymnastics' in a clearing and then research. He conducted a census, carried around in a chair by a retinue of boys, much like the local chiefs, who sat raised up so that their people did not need to crawl as they passed them. It gave him, he admitted, a 'delightful feeling that now I alone am master of this village'. Being back in New Guinea, ferried in a canoe through crystal waters as his Motuan oarsmen sang, was wonderful: 'I imagined you [Elsie] with me today, sailing over the pale green waters ... the whole time I "talked to you" and formulated things for you.' He was deeply moved by the beauty of the lagoon on which he lived, the jagged mountains and green forests with their intoxicating scents of frangipani and tuberose, the jewel-like colours of the landscape, ocean and sky, phosphorescence, moonlight and the stars at night; overcome by a

Malinowski learning about the kula, *an ancient trading network, from his Trobriand hosts, in 1917.*

'*joie de vivre tropicale*, something like being drunk on strong wine, at once oppressive and stimulating'.

His equipment, especially photographic, gave him endless trouble. There was never enough clean water for developing, fading light frustrated him, occasionally he realised he'd forgotten to load the camera with film. When he found the 'natives' infuriating, he longed for 'civilisation'. He was both attracted and repelled by the golden skin and relative uninhibitedness of the women whose sexual habits he was examining in the name of science; writing in his diary provided him with a channel for repressing his 'lecherous thoughts' and dreams. Weeks might pass without him feeling he had achieved anything or engaged meaningfully with his subjects; he berated himself for sluggishness, moral weakness, greed and ambition, for smoking, lustful impulses and addiction to novels.

Endlessly examining his health (and dosing himself generously with arsenic and quinine), frequently lonely to the point of despair, he was plagued by what he called the 'considerable psychic strain' involved in realising an alternative vision of the world. He was engaged in the most profound of existential enquiries. 'What is the deepest essence of my investigations?' he asked himself. 'To discover what are his [the Trobriander's] main passions, the motives for his conduct, his aims ... His essential, deepest way of thinking. At this point we are confronted with our own problems: what is essential in ourselves?'

For much of the time Malinowski spent in the Trobriands, he was consumed with distant personal problems, exacerbated by the difficulty of communication (even worse during wartime). Letters took a month at least just to reach Australia; to and from Europe might take a year. He missed Elsie and thought constantly about her, spending one day off drawing up a chronology of their meetings, and about his hopes of what they could be to one another. If she had 'faith in my heroism', he was determined to live up to her ideals.

Then there was the engagement to Nina Stirling, which he needed to break off before he could fully commit to Elsie, of whom he now thought as his future wife. Racked with guilt when he heard that Nina was ill, he couldn't bear to read her letters, putting off making a definitive break with her and hating himself for his hesitation. Towards the end of his stay in the Trobriands, Nina discovered she had a rival and informed Elsie directly that she and Malinowski could not have been on more intimate terms if they had been married. It is clear from the diary that he and Elsie had also been on those intimate terms – he remembered her in his rooms in a red kimono – and he had to acknowledge that both women had suffered from their association with him. He felt himself to be not just weak but wicked.

A putteed and pith-helmeted Malinowski examining the necklaces of Trobriand women. He was plagued by 'lecherous thoughts' during his time in the Tropics.

Work became his solace and his escape. He dreamed of coming home 'laden with materials as a camel', of being recognised as 'an eminent Polish scholar'. At times he was suffused with contentment at the peacefulness of his life and his absorption in work: 'I am completely under the spell of the tropics as well as under the spell of this life and my work. For nothing in the world would I read trashy novels, and I think with pity on people who keep taking medicine all the time!'

But after six months away, in June 1918, he heard that his beloved mother had died. He was overcome with a grief bound up with a desperate sense of his own inadequacies, longing, he wrote, for the 'shallow optimism of religious belief'. At night he wept on the beach, in the morning he woke from dreams suffused with his mother's presence, 'drowned with sadness'. The world had lost its colour and meaning and was full only of remorse. The following month he wrote the final sentence of his diary: 'Truly I lack real character.'

His diary was a confessional, never intended for publication,* but Malinowski apparently did view it as complementing *The Argonauts of the Western Pacific* (1922), his scholarly account of his months in the Trobriands, an admission of the idea that ethnography is informed implicitly by autobiography as much as explicitly by theory and method. Not until the diary was posthumously published in 1967 would this insight truly exert

---

* It was published by his widow, Valetta Swann, who did not seek the permission of Malinowski's three daughters.

its influence over the discipline, controversially illuminating 'something of the process by which the modern anthropological viewpoint was achieved'; for the moment, Malinowski's struggles and limitations remained private.

Ironically, it's not the parts of his diary in which he wrestles with his conscience, or examines his character and finds it wanting, that shock today in our world of mental health awareness and the normalisation of analysis, but the attitudes and language he uses without thinking. 'I see the life of the natives as utterly devoid of interest or importance, something as remote from me as the life of a dog,' he wrote, casually referring to 'savages', 'natives', 'blacks' and 'niggs'. He could use the word 'civilisation' without irony at a time when the 'civilised' world was engaged in a brutal war that would see the slaughter of hundreds of thousands, while condemning the 'savages' hosting him for petty thievery and lasciviousness.

And yet, while he privately thought like this, he was also formulating a new way of viewing the 'savages' among whom he'd lived that would transform the way so-called civilised people viewed the inhabitants of the rest of the world. *The Argonauts of the Western Pacific* focused on the *kula*, an ancient trade and gift-giving network spanning the Trobriand Islands that lay 'on the borderland between the commercial and the ceremonial'. In elaborate feasts and rituals, islanders exchanged beads and shell armlets (for men) and necklaces (for women), which were highly valued and eventually exchanged again. Malinowski showed that trade and enterprise were inextricably linked with magic and status, forming 'one inseparable whole'; over the decades to come, his findings would inspire Marcel Mauss in Paris and lead to important anthropological work on gift-giving and reciprocity. In the words of Edmund Leach, one of his students, he presented the Trobrianders 'as both unique and universal': different from the implicit us – his civilised readership – but also, crucially and really for the first time, the same.

He concluded the book with a plea for understanding that encapsulated all that anthropology, as he saw it, was trying to do. Although his words reveal the unconscious prejudice of his day, they also underline his commitment to seeing other people through fresher, clearer eyes: 'In grasping the essential outlook of others, with the reverence and real understanding, due even to savages, we cannot help widening our own. We cannot possibly reach the final Socratic wisdom of knowing ourselves if we never leave the narrow confinement of the customs, beliefs and prejudices into which every man is born ... The Science of Man, in its most refined and deepest version should lead us to such knowledge and to tolerance and generosity, based on the understanding of other men's point of view.'

With *Argonauts*, according to Adam Kuper, Malinowski was credited with bringing 'theory into the field'. Laying out the principles of his method, he followed Rivers by emphasising the scientific aims of any fieldwork and the active pursuit of ethnographic evidence, as well as the importance of living immersed in the society being studied and ideally speaking the language. The gloss he added was the idea that if a researcher was living 'like a native', she would no longer excite attention or provoke self-consciousness among her companions and could *observe* them almost as a friend, through *participation* in their daily life, acquiring '"the feeling" for native good and bad manners'.

What Malinowski also uniquely contributed was the sense of himself as the protagonist of his work – the anthropologist as hero. His native subjects were treated with benign disdain and the knowledge of other Europeans long living in the region was often summarily dismissed. Despite his lack of interest in their opinions on local matters, he was quite happy, according to his diary, to escape work in the company of the various European traders and missionaries who lived in and around Kiriwina, though needless to say he did not mention them in *Argonauts*. As the anthropologist Michael Young, author of several books on Malinowski, observes, although he sought man 'in the round and not in the flat' (quoting Sir James Frazer, from his introduction to *Argonauts*), he is the one who emerges in his writing as round, vital, fully fleshed; his informants and guides do little more than serve as relief.

He could be his own worst enemy, as patronising to his indigenous hosts as he was pompous and narcissistic. This comes from his introduction to the 1935 *Coral Gardens and their Magic*, his own favourite work about the agriculture of the Trobriand Islands: 'Once again I have to make my appearance as a chronicler and spokesman of the Trobrianders, the Melanesian

community so small and lowly as to appear almost negligible –
a few thousand "savages", practically naked, scattered over a
small flat archipelago of dead coral – and yet for so many reasons
so important to the student of primitive humanity.' It was this
attitude, as demonstrated in his diary decades later, that prompted
the anthropologist Clifford Geertz to comment that Malinowski's
success as a fieldworker was 'a triumph of sheer industry over
inadequate empathy'.

Malinowski returned to Melbourne, as laden down with notes
as he had hoped, untangled his private life and married Elsie in
1919, just as the war ended and they could return to Europe. As
he had intended while he drew his 'last ethnological escapade'
to a conclusion, the material he had gathered in New Guinea
would see him through his career: *Argonauts* in 1922, *Crime and
Custom in Savage Society* in 1926, *The Sexual Life of Savages in
North-Western Melanesia: An Ethnographic Account of Courtship,
Marriage, and Family Life Among the Natives of the Trobriand
Islands, British New Guinea* in 1929 and *Coral Gardens and their
Magic* in 1935. By 1922, the year *Argonauts* was published and
declared a masterpiece by his peers, he was living in London with
Elsie and lecturing at LSE alongside Charles Seligman and Edvard
Westermarck. He became reader in 1924 and professor in 1927.

As a teacher, Malinowski was as controversial as he was
charismatic, given to making provocative jokes about the discipline
to which he had devoted his life. He liked to quote in answer to
the question about the customs and manners of the people of

New Guinea, 'Customs none, manners beastly', and he defined anthropology as 'the study of rude man by rude men'. (Margaret Mead would cap this, defining anthropology, perhaps particularly Malinowski's, as 'the study of man, embracing women'.) His lectures were legendary. 'With a suave question, a caustic word or a flash of wit, he would expose a fallacy, probe for further explanation and after inviting opinion from all sides, he would draw together the threads in a masterly way,' wrote George Stocking, 'lifting the whole discussion to a higher theoretical level.' In his seminars, remembered one student, 'he made *us* think'.

Hortense Powdermaker (whose most celebrated book was the 1950 anthropological study, *Hollywood, the Dream Factory*) was taught by Malinowski at LSE in the 1920s and, despite her awareness of his faults, found him stimulating and exciting. He was, she wrote, a man of paradox: 'kind and helpful as well as cruel and sarcastic', belligerent and boastful but also dynamic and perceptive. He demanded loyalty but adored irreverence, and made of his favourites a family in much the way Franz Boas did at Columbia, revelling in what he liked to call their 'pedestalisation' of him. Chosen students were invited to take dictation or even join him on working holidays at his chalet in the Dolomites, where Malinowski would lie in the sun on the balcony, nude but for a green eyeshade and a tanning solution of salt and iodine. (His name features on the membership list of interwar London's most glamorous and bohemian nightclub, the Gargoyle; he was noted for his tango and a wild Charleston.)

When Ruth Benedict, who worked with Franz Boas, met Malinowski in New York, she wrote excitedly to Margaret Mead about his 'quick imagination and ... by-play of mind ... It's intriguing to find an intelligent person discovering with such force the things we've been brought up on with our mother's – or Papa Franz's! – milk.' But not everyone was susceptible to his spell. From Paris, Marcel Mauss wrote to sympathise with Alfred Radcliffe-Brown about Malinowski's 'despotism'. 'Malinowski is forever engaged in two favourite pastimes,' observed Benedict's colleague, another Boasian, Robert Lowie. 'Either he is battering down wide-open doors; or he is petulantly deriding work that does not personally attract him.'

The work that did interest him was methodology, the relationship between research, observation, hypothesis and analysis. As he wrote in *Argonauts*, 'Not even the most intelligent native has any clear idea of the Kula as a big, organised social construction, still less of its sociological function and implications ... The integration of all the details observed, the achievement of a sociological synthesis of all the various, relevant symptoms, is the task of the Ethnographer ... the Ethnographer has to *construct* the picture of the big institution, very much as the physicist constructs his theory from the experimental data, which always have been within reach of everybody, but needed a consistent interpretation.'

Functionalism is the academic concept with which Malinowski has been most associated. He believed that society functioned to meet the needs of individuals and that it was (in the words of David Shankland) 'possible to see all social life in terms of an expanding series of interwoven events which give meaning to and influence each other'. As his student Audrey Richards put it with characteristic clarity, 'rites, beliefs and customs, however extraordinary they appear to an observer, actually fill "needs", biological, psychological, and social'. When Edvard Westermarck gave the Huxley Memorial Lecture in 1936, he quoted Malinowski on functionalism, by then the guiding theory behind anthropology. Functionalism, wrote Malinowski, 'insists on the complexity of sociological facts, on the concatenation of various often apparently contradictory elements within one belief or conviction; on the dynamic working of such a conviction on the social system; and the expression of social attitudes or beliefs in traditionally standardised behaviour'.

He believed that societies were made up of and for individuals. Looking at human society as a body, if the skeleton is society's institutions, and flesh and blood people's daily life and ordinary behaviour, the intangible but guiding soul is their words, stories and beliefs. He also deemed it crucial to observe the differences between what people said and what they did. Edmund Leach learned from him 'that [society's] rules were there to be broken, and that individuals pursue their own interests regardless [of those rules]'. The family, therefore, rather than the clan, was the primary form of social relationship; most people, Malinowski observed, would pay 'lip service to the public morality of the clan, while doing their best to further the private interest of their families' – like the unspoken 'vote pink, live blue' credo of the champagne socialist.

The disciplinary movement away from its primary emphasis on religion and magic towards kinship, in Rivers' and Radcliffe-Brown's work, continued with Malinowski towards economics and politics. Whereas the first anthropologists – Tylor and Frazer pre-eminent among them – had focused, from their libraries, on the origins of civilised man, Malinowski's generation, out in the field, would turn their attention to a close observation of what was, not what had been. With Radcliffe-Brown, this resulted in a strictly ahistorical approach; he was interested not in the past but in what was happening in front of him. Malinowski concentrated on the interconnected relevance of everything, the 'minute contextualisation' of details gathered in the field.

Malinowski also understood that in order for anthropology to appeal to the broader public, it had to address issues and themes that had contemporary relevance. This meant moving away from archaeology and prehistory – with which the first anthropologists had been closely associated – and towards interdisciplinary human sciences. The problems of the post-war, post-Victorian world, edging towards a post-colonial future, could be seen in a new light through an anthropological lens, asking questions always in terms of what humans have in common, not what separates us. Although anthropology relied for part of its appeal upon the exotic, what made it fascinating to a non-academic audience was its ability to make that exotic seem familiar. Malinowski's thinking, wrote

Powdermaker, was never restricted to 'savages': 'In lectures he often gave comparative examples from modern civilisations, and in conversations he made amusing allusions to contemporary British tribal rites and customs. He was a cultivated humanist as well as a scientist, and ... he saw no conflict between the two views.'

His populist approach provoked criticism – this was academia, after all – with Rafael Karsten pointedly enquiring, 'Why would you, for example, call a work *Crime and Custom in Savage Society* when all you are talking about are the Trobriands?', but more readers were in agreement with Havelock Ellis, who commented of Malinowski's work, 'we find not merely that in this field the savage man is very like the civilised man, with the like vices and virtues under different forms, but we may even find that in some respects the savage has here reached a finer degree of civilisation than the civilised man. The comparisons we can thus make furnish suggestions even for the critical study of our own social life.'

Ellis's words were part of his introduction to *The Sexual Life of Savages*, Malinowski's most notorious book, which was wrapped in brown paper for the first few decades of its published existence. As ever, Malinowski made full use of the curious customs he observed to pull in the general reader. He described *mitakuku*, the tender biting off of a lover's eyelashes: 'I was never quite able to grasp either the mechanism or the sensuous value of this caress. I have no doubt, however, as to its reality, for I have not seen one boy or girl in the Trobriands with the long eyelashes to which they are entitled by nature.'

> *Bamasisi deli   Dabugera;   bayobobu,*
> I sleep together   Dabugera;   I embrace,
> *bavakayla   bavayauli.   Tanunu dubilibaloda,*
> I hug all length, I rub noses.   We suck lower lips ours,
> *pela   bi'ulugwalayda;   mayela   tanunu;*
> because we feel excited;   tongue his we suck;
> *tagadi   kabulula;   tagadi   kala gabula;   tagadi*
> we bite   nose his;   we bite   his   chin;   we bite
> *kimwala;   takabi   posigala,*
> jaw (cheek) his;   we take hold (caress)   armpit his,
> *visiyala.   Bilivala   minana: "O didakwani,*
> groin his.   She says   this woman: "O it itches,
> *lubaygu,   senela;   kworikikila*
> lover mine,   very much indeed;   rub and push
> *tuvayla,   bilukwali   wowogu—*
> again,   it feels pleasant   body mine—
>
> *kwopinaviyaka,   nanakwa   bipipisi*
> do it vigorously,   quick   (so that)   it squirts
> *momona:—   kwalimtumutu tuvayla bilukwali*
> sexual fluid:—   tread   again   it feels pleasant
> *wowogu."*
> body mine."

A lexicon from The Sexual Life of Savages *(1929) which, for some decades, was considered so obscene that it was sold wrapped in brown paper.*

Although he recorded sexual freedoms for young people unthinkable in European society, he argued that stable unions usually ensued – just as Westermarck had theorised, this was not the pattern of primitive promiscuity once thought to have marked early human culture. A gradual introduction to sex prepared Trobriand youths for sexually satisfying but realistic partnerships; although adultery, jealousy and divorce did occur, infidelity did not always lead to a marriage breaking up. He observed matriliny but argued that it did not necessarily mean matriarchy (a challenging idea for most Western thinkers of the 1920s, however progressive).

Despite his efforts to be objective, in some areas Malinowski showed himself to be very much a man of his times. There are only five citations of homosexuality in a book of nearly 500 pages and the index entry reads, 'Homosexuality: contempt for and repugnance to; unnatural conditions conducive to'. He was, however, fascinated by childhood sexuality, writing that children began their 'real sexual life' at about seven for girls and eleven for boys. Evidently providing inspiration for the children of Aldous Huxley's 1932 dystopia, *Brave New World*, he described play as highly sexualised, with older children often initiating younger in genital manipulation and oral stimulation 'long before they are able really to carry out the act of sex'.

He also sought to demonstrate that the male Oedipus complex was not found in the Trobriands, suggesting that in a matrilineal society the 'repressed desire' described by Sigmund Freud is not to kill the father and marry the mother, but to kill the maternal uncle and marry the sister. 'Malinowski was an outstanding anthropologist because he usually recognised an important problem, and having recognised it, he almost always formulated the relevant research question,' wrote Melford Spiro fifty years after *Sexual Life* came out, challenging its conclusions.

According to Spiro, the question Malinowski asked was 'Do the conflicts, passions and attachment within the family vary with its constitution, or do they remain the same throughout humanity?' In Melanesia, the supreme taboo, as Malinowski observed, was brother–sister incest, which again fitted in with Westermarck's theory of reverse sexual imprinting.

In the end, Malinowski found that contemporary Western society's obsession with the sexual lives of others said more about itself than it did about those others. 'The so-called savage has always been a plaything to civilised man,' he observed. 'Savagery has been, for the reading public of the last three centuries, a reservoir of unexpected possibilities in human nature; and the savage has had to adorn this or that *a priori* hypothesis by becoming cruel or noble, licentious or chaste, cannibalistic or humane according to what suited the observer or the theory.' He concluded that the morality of the average Trobriander was 'more or less on a level with that of the average European – that is if the customs of the latter were as frankly described as those of the Trobriander'.

His academic work on sexual practices in the 'primitive' world gave him an interest in sex education in the 'civilised' world to which he belonged and he began campaigning alongside Marie Stopes and Margaret Sanger for birth control, which could not be legally prescribed by a doctor in the United States until 1917. (Both Sanger and Stopes had eugenicist views – Stopes's contraceptive cap was marketed under the name Pro-Race.) He became passionately interested in the social and moral benefits to which he felt anthropology could contribute. Birth control was one; improving the lives of the 'natives' with whom he worked was another. In 1926, introducing *Crime and Custom*, he lamented the civilised world's neglect of 'the study of the rapidly vanishing savage races' that it was destroying. 'The task [anthropology] is not only of high scientific and cultural importance, but also not devoid of considerable practical value, in that it can help the white man to govern, exploit and "improve" the native with less pernicious results to the latter.' Three years later, in an article entitled 'Practical Anthropology', he called for the lessons and principles of anthropology, practical, modern, rigorous and unbiased, to be applied in colonial states.

Where Radcliffe-Brown, often seen as his academic and theoretical counterpoint, was scrupulously apolitical, Malinowski admitted to 'a deepening personal radical commitment'. Even in his Trobriand diary, littered as it is with unthinking racial slurs, he expressed disgust at white, colonial attitudes of 'superiority'. By 1945, his thinking had crystallised. He believed the anthropologist had a duty 'to be a fair and true interpreter of the Native ... Shall we, therefore, mix politics with science? In one way, decidedly "yes" ... He ought to make clear to traders, missionaries, and exploiters what the Natives really need and where they suffer most

---

* In 1937, Malinowski, who insisted he enjoyed his own 'caricature silhouette' by the ever-critical Robert Lowie, was amused to note his rival Radcliffe-Brown's pedestal shaking 'under the blows of the chisel': more usually it was his pedestal upon which the blows rained down.

under the pressure of European interference. There is no doubt that the destiny of indigenous races has been tragic in the process of contact with European invasion.'

Later anthropologists and historians would charge Malinowski with being solipsistic and hypocritical – although the fact that even decades after his death his work was still controversial is a measure of its weight. In claiming usefulness to colonial administrations, his critics pointed out, he was leaving himself and his colleagues free at once to sympathise with the natives among whom they lived and to pursue their own academic interests. Wendy James, writing in the early 1970s, saw 'the colonial anthropologist as a frustrated radical: and his claims to scientific status, the separation of his work from any apparent moral or political views, and the avowals of its practical usefulness, as largely determined by the need to make a convincing bid for the survival and expansion of his subject'. In fact, the demise of colonialism would deprive the European discipline, at any rate, of much of its impetus. As Helen Tilley and Robert Gordon succinctly observed in 2007, 'anthropology needed empires far more than empires needed anthropologists'.

But all this, and the furore over Malinowski's diary, was still to come. When he died in 1942, he was a 'pedestalised' professor at Yale, where he had been living since 1939 with his second wife, the English painter Valetta Swann; his beloved and understanding Elsie had died in 1935. The great successes of his career had been founded upon the year or so he spent in the Trobriand Islands, so perhaps the islanders should have the last word. While both the Toda studied by William Rivers and the Inuit among whom Franz Boas lived believed the ethnologists to have brought bad luck to their people, and no doubt waved them off with relief, the people of New Guinea chose to commemorate Bronislaw Malinowski with an appropriately phallic memorial, a stalactite broken off from the roof of a coral cave, placed on the site where his tent stood in their capital, Kiriwina (now replaced with a plaque). The three nicknames they gave him also survive, revealing that they had observed him as closely as he observed them: Topwegigila, the man with baggy shorts (those 'knickers' expensively made for him in London in 1914); Tololibogwa, the collector of stories; and Tosemwana, the show-off.

CHAPTER SIX

The Academy

Franz Boas at Columbia University, 1899–1942

————•————

In May 1906, Franz Boas, head of the anthropology department at Columbia University in New York, addressed the twenty-one graduating students of Atlanta University, one of the early African American colleges, now called Clark Atlanta University. He had been invited at the behest of W. E. B. du Bois, the pioneering African American academic and president of the university. It was a critical moment in race relations and du Bois (a professor of sociology as well as history) hoped an anthropological approach – the latest thing in the world of academia – might inspire his graduates as they prepared to begin their adult lives in a divided and challenging United States.

Over the previous decades, Southern legislatures had worked effectively to disenfranchise black Americans, who had been granted the right to vote at the end of the Civil War, and bed in legal segregation. With 302 lynchings between 1901 and 1931, Georgia was the state with the highest number of incidents during this period, and in the city of Atlanta alone, 64 black men were murdered by white vigilantes in 1906, the year Boas spoke at the university. After what were euphemistically called 'race riots' there later that year, 'the sidewalks ran red with the blood of dead and dying Negroes'.

In 1909, du Bois would be one of the founding members of the National Association for the Advancement of Colored People, an organisation created to combat the rising tide of violence, injustice and discrimination in the United States. The NAACP sought to counter the powerful influence of the civil rights activist and former slave Booker T. Washington, who advocated 'accommodation' with the white South – which despite its intentions in effect meant tacit obedience to white rules and acceptance of second-class status.

Du Bois knew that Boas had progressive views on race but he was still dumbfounded by the speech Boas gave on that summer's day. 'You need not be ashamed of your African past,' he told the audience before admiringly recounting 'the history of black kingdoms south of the Sahara for a thousand years', detailing their artistic and technological sophistication and their intricate forms of government. He urged his young listeners to have pride in their forebears and confidence in their potential. 'I was too astonished to speak,' remembered du Bois, who had received his second degree, in history, *cum laude*, from Harvard. 'All of this I had never heard …'

Boas concluded by comparing the situation of black people in America with that of Jewish people in Europe. 'Remember that

Franz Boas in 1920. The facial scars he received as duelling student can just be seen, though it is the intense gaze that strikes the viewer.

in every single case in history the process of adaptation has been one of exceeding slowness,' he exhorted them. 'Do not look for the impossible, but do not let your path deviate from the quiet and steadfast insistence on full opportunities for your powers.' His Jewish background and his youth in Germany had made him particularly sensitive to prejudice in any form – the scars of that sensitivity were still visible on his face – and his career as the father of American anthropology would be defined by an assault on the idea (in the words of Marshall Hyatt) 'that cultural differences were racially determined'.

The emerging American school of anthropology was dominated by disputes about racial politics in a way that the British school at this stage was not, despite their colonies, because 10 per cent of the United States population were impoverished African Americans bearing the bitter legacy of slavery, as well as Native American people who had been murdered, brutalised and marginalised since European settlers arrived three centuries earlier. Indigenous people were considered neither citizens nor foreigners but 'wards' of the federal government, their legal status reflecting the view that they were not quite Americans – perhaps not even quite people.

Lothrop Stoddard was the sort of public intellectual against whom Boas needed to set out his stall, so widely known that the brutish Tom Buchanan mistakenly eulogised him as 'this man Goddard' in *The Great Gatsby*. He was a historian, educated at Harvard, whose bestselling 1920 book *The Rising Tide of Color Against White World Supremacy* expounded his eugenicist views. (He was also, like Bronislaw Malinowski, a committed advocate of birth control.) In 1929, he would make the mistake of agreeing to debate white supremacy with du Bois in Chicago. Du Bois, who answered the questions 'Shall the Negro be encouraged to seek cultural equality? Has the Negro the same intellectual possibilities as other races?' in the affirmative, gambled that Stoddard would look an idiot defending his indefensible views. He was right: the 5,000-strong audience cheered du Bois and laughed at Stoddard, who might as well have been wearing his silly KKK hood.

But ridiculing thinkers like Stoddard – or even allowing them to make themselves ridiculous – diminishes the real threat they pose. Scientific racism was an absolutely serious academic theory in the late nineteenth and early twentieth centuries, received thinking in the highest intellectual circles; Boas, who believed scientific truth would 'blaze a path to a better world', saw it as his anthropological mission to refute it. He was not alone and he was not the first – in his 1901 presidential address to the American Folklore Society, Frank Russell had cautiously praised ethnologists for learning both to 'take a more modest view of the virtues of the Caucasian' and 'to credit the savage and barbarian with many praiseworthy qualities' – but Boas expressed his views with greater force and clarity, becoming the torchbearer for an entirely new way of considering race.

The foundation of Boas's approach was the repudiation of the nineteenth-century evolutionary theory that there was a rough hierarchy of human races from apes up to the white man (cue angels singing) through the Negro and the Australian Aborigine, and the stages of this hierarchy – which peoples were more primitive and therefore more deserving of ethnological study – were hotly debated. Though views were changing in Europe, influenced by the theories of diffusion that were so important to William Rivers, in the United States, before Boas took up his position at Columbia, anthropological research was still dominated by social evolution. Races were seen as both culturally and biologically different. Boas discredited these theories so fundamentally that by the time he died, no anthropologist would have dreamed of suggesting that race was a physical distinction. In a series of articles and books, beginning with his landmark anthropometric study *Changes in the Bodily Form of Descendants of Immigrants* (1911), he countered the false science that sought to prove that some races were superior to or even fundamentally different from others: 'It is quite impossible to say that, because some physical function, let us say the heartbeat, has a certain measure, the individual must be White or Negro – for the same rates are found in both races.'

Having contributed to the 1911 *Handbook of Indian Languages*, Boas observed that the native tribes of North America were very far from uniform, as popularly imagined, but instead as different from each other as they were from other 'races'. Attempts to classify race through anatomy, geography or language were all artificially imposed, he argued, and therefore doomed to failure because cultures were so profoundly but variably intermingled through diffusion. In the *Handbook* and *Race, Language and Culture*, published in 1940, he used examples from Europe, where waves of migration over millennia into Spain and Britain had produced inextricably 'mixed' populations, and China, where a multitude of tongues revealed a multiplicity of peoples despite a predominant racial 'type'.

Environment, rather than race, was the primary shaper of behaviour. He demonstrated that variations in the skull shape of immigrants correlated with how long they had been living in the United States, even over a single generation; the longer they had been there, the 'rounder' their heads were. One student, Maurice Fishberg, showed in minute studies of hair and eye colour, height, skull size and nose length of Jews in various countries from North Africa to eastern Europe that there is 'more difference between the Caucasian and North African Jews than there is between the Russians and Germans'. When faced with IQ tests showing that northern Europeans were more intelligent than southern Europeans, and Europeans as a whole were more intelligent than Negroes, Boas countered that nutrition and education accounted for the differences. There was a marked contrast between rural and urban populations: the longer a black man lived in a city, the higher his scores on these tests became.

Wherever he could, he supported thinkers challenging racism. In 1911, he contributed the introduction to *Half a Man: The Status of the Negro in New York*, researched and written by Mary Ovington, a friend and supporter of W. E. B. du Bois and co-founder of the NAACP. Her study, he wrote, was 'a refutation of the claims that the Negro has equal opportunity with the whites, and that his failure to advance more rapidly than he has is due to innate inability'.

Critics have called Boas a reformer rather than a radical; despite his progressive views he remained, according to Vernon J. Williams, 'a prisoner of his times'. His attitude to 'interbreeding' – a much-debated, highly controversial topic at the time – is a case in point. Unusually, he did not think it caused 'degeneracy' but rather hoped the subsequent blurring of skin colour would lead white Americans to accept African Americans as equals – although, illogically, he favoured black women having children with white men, believing their offspring would be less 'dark' than the children of black men and white women. In 1915, the craniologist Robert Shufeldt sent his book *America's Greatest Problem: The Negro* to Boas for his comments. Boas replied, 'I am afraid that you would not wish to include an expression of my opinion in your book because I am not at all convinced that the miscegenation of the races is a bad thing ... If mixed race children fall short of "the mark"', he added pointedly, it was because in contemporary America they were forced to 'grow up under very unfavourable circumstances'.

Later criticism of Boas, notably by William Willis, who received his doctorate from Columbia in 1955, more than a decade after Boas died, identified the 'double standard [that] clearly pervaded field procedures', with bullying, paid-for information and underhanded tactics used to research black and Indian people that would have been unthinkable for white subjects, including measuring people and grave-robbing. Willis was the first of many within the discipline to argue that Boas did nothing to reform 'the basic function of anthropology: the improvement of white society and white people'.

It was undeniably true that Native Americans and later African Americans served much as the various peoples of the British Empire did for the anthropologists of Cambridge and LSE, as little more than the obvious (albeit fascinating) subjects of ethnographic fieldwork. Boas took up his position at Columbia in the shadow of the inspirational one-armed geologist and explorer John Wesley Powell, who set up and ran the Bureau of Ethnology (from 1897 the Bureau of American Ethnology, or BAE) in the 1880s and 1890s. The BAE's stated mission was the transfer of all material relating to indigenous people – American Indians, as they were known – from the Department of the Interior to the Smithsonian Institution, but it became a much broader anthropological research project.

Powell, a generation older than Boas, was an evolutionist who saw the indigenous people his department studied as primitives whose lives would be improved and enlightened by contact with the 'civilised' world. Implicit in this, however – as in the insights of William Rivers and Bronislaw Malinowski and their colleagues in the field – was the understanding that contact with white men would destroy the languages, customs, arts and beliefs of these people. The unique task of American ethnologists, as Powell saw it, was to record the savage or barbaric cultures that had preceded his in the United States and which would soon, inevitably, disappear entirely in the name of progress, a process known as salvage ethnology.

Boas learned a great deal from his work with Powell and the BAE, and continued its research into Native American societies, but he moved away from Powell's tacit acceptance of the extermination of native 'races' on the grounds that they were remnants of savage cultures, mere precursors to civilised man. Resisting attempts to view the cultural practices of 'savages' as imperfect approximations of European ones, he argued (in the words of George Stocking) that other peoples were 'quite differently constituted cultural categories that were at best problematically commensurable to a Eurocentric evolutionary standard'. As the theoretical framework of his anthropological school shifted from cultural evolution to historical particularism, he implored his students not to judge other cultures by comparing them to their own. He liked to quote Immanuel Kant, whose work had made such an impression on him in the Arctic: 'Forget about pigeonholes, they're all in your head. Give your attention to the pigeons; the things you can actually see and touch. They are the only reality.'

Where Boas's Department of Anthropology at Columbia differed from any previous American ethnological endeavour was in its professionalism. Under Boas, anthropology would become a discipline; bringing it out of the realm of the amateur and into academia gave it an entirely new quality. It was his desire to establish 'a well organised school of anthropology ... one of the fundamental needs of our science', he wrote to the archaeologist and anthropologist Zelia Nuttall (a friend since his Chicago days) in 1901, and he believed that it would be 'of advantage to American anthropology if I can retain a certain amount of control'. Recognising that this ambition represented an implied critique of any other approach, one of his students and colleagues, Pliny Earle Goddard, admitted that Boas's primary goal was 'to build up the "Boas school"'.

The group that coalesced at Columbia in the first quarter of the twentieth century around the sprightly figure of Professor Boas was like a family. Margaret Mead described him in 1922 with a slight, frail body and large head, his face marked by both the scars of his old duels and a drooping eye; another student thought he was like a fighting cock, small but plucky. Their thrilling 'new discipline, with little history and an uncertain future' attracted maverick minds who wanted to write their own rules, guided by their 'idealistic slave-driver', their 'beloved hero', their Papa Franz. They were engaged less 'in learning a body of received wisdom than in creating one'.

Boas stood as if on Parnassus, pointing out to his followers 'the land below, its shadowed parts and its sunny places alike virgin to the ethnologist. Virgin but fleeting – this was the urgency and the poetry of Boas's message.' Thus wrote Theodora, the widow of the first of Boas's student colleagues, Alfred Kroeber,* who in 1901, after receiving the first doctorate in anthropology from Columbia, established the anthropology department at the University of California in Berkeley. 'Everywhere there were to be discovered Ways of Life, many many ways. There were gods and created worlds unlike other gods and worlds ... all soon to be forever lost, part of the human condition, part of the beautiful heartbreaking history of man,' Theodora continued. 'To the field then! With notebook and pencil, record, record, record. Rescue from historylessness all languages still living, all cultures. Each

---

* Kroeber and Theodora's daughter is the novelist Ursula Le Guin.

is precious, unique, irreplaceable, a people's ultimate expression and identity, which being lost, the world is made poorer.'

It was a small world, made more intense by time spent in the field and the sense of mission its inhabitants were encouraged to feel about their work, seeing it as their intellectual duty 'to exhibit the rich diversity of human cultures'. They fell in and out of love with one another – Kroeber married one of his students when his first wife died; another of his students, Gladys Reichard, lived with (but didn't marry) Boas's student, Pliny Earle Goddard; lanky Edward Sapir wrote poetry with Ruth Benedict and pursued Margaret Mead when his first wife died. Many came from similar European intellectual backgrounds, often Jewish, which meant they directly shared Boas's outsider gaze and his concerns about racially based science. As the Vienna-born Robert Lowie put it, his 'marginality to typical American culture' made him especially susceptible to the deep impressions made upon him by other marginal people. Columbia was one of the few elite universities at this time not to unofficially limit their intake of Jewish students.

Boas was an all-consuming leader, giving generously but insistent upon rigour and fealty. 'He seemed to personify the very spirit of science, and with his high seriousness – unsurpassed by any investigator I have known in any sphere – he communicated something of that spirit to others,' wrote Lowie after Boas's death. Each of his students was required to sit notoriously difficult courses on statistical theory and Native American languages, and he thought nothing of assigning work in an unfamiliar language.

That interdisciplinary fluidity that had marked his Arctic expedition in the 1880s was a hallmark of his academic work, too, spanning 'history (including ethnohistory), linguistics, literature, folklore, museum studies, philosophy, science studies, politics, law, education and psychology' as well as biology, geography and archaeology. At the same time, in flashes, he could universalise his subject, observing in a discussion on taboo and magic that as a student he would have challenged to a duel anyone who had destroyed a photo of him; his scarred face was a measure of his candour.

But some of his colleagues – as evidenced in the correspondence between Sapir and Lowie in the 1920s – most of whom had been his students, privately felt that daring to disagree with him would create a 'family quarrel', writing to each other about both the impossibility of falling out with him and the necessity 'to break away from him, and not always gently' in order to reach intellectual maturity. Although cultural pluralism was the ideal Boas espoused, in practice he could be patriarchal and rigidly intolerant of dissent. As an outsider, Hortense Powdermaker found the cult of personality he encouraged troubling.

Rising above these personal entanglements, Boas steered students into the areas (theoretical and geographical) that he felt needed attention, within his 'four fields' system of physical anthropology, linguistics, archaeology and cultural anthropology. Kroeber, his 'elder son', took his message to the West Coast, along with Lowie, who joined him there in 1917 and worked predominantly on social structure; under Boas's aegis, Clark Wissler replaced him as curator of ethnology at the American Museum of Natural History, concentrating on material culture; Paul Radin studied anthropology with relation to history and philosophy.

Boas emphasised the importance of language's overlap with culture, especially as its forms were nearly always unconscious. Using Latin and English grammar, which are far simpler than most primitive languages, he demonstrated that cultural phenomena didn't always move from simplicity to complexity. He insisted that languages be analysed on their own terms, not according to how they fitted into Indo-European categories; with his encouragement, Sapir focused on linguistics, classifying the myriad of indigenous American languages.

While Boas (like the forward-thinking discipline of anthropology as a whole) was ahead of his time in welcoming female students, he retained a rather nineteenth-century view of them – women, he thought, were more patient, more compassionate and more creative than men – which meant he steered them towards what he saw as their areas of strength: folklore, art and religion. It went without saying at this time that women were best suited to investigating domestic culture and anything to do with children and families.

It was an irony that the woman who worked with him most closely throughout his career was neither his student, nor remotely dependent upon him. Elsie Clews Parsons was a free-thinking heiress who had received her doctorate in sociology from Columbia just after Boas arrived there, in 1899. As a folklorist, she was a skilled and influential fieldworker in her own right, spending decades studying both Native American and African American culture, but arguably her greatest contribution to anthropology as a discipline was her financial support of Boas. She paid for his private secretary and, crucially, funded the fieldwork of many of his students and the publication of their findings, allowing them the luxury of working unconstrained by external requirements.

After his own revelatory months in the Canadian Arctic, Boas required each of his doctoral students to perform an extended period of research in the field. He stressed the fundamental importance of empirical study, speaking the language of the informants directly and interpreting data from their viewpoint. He outlined the four principles of his method: participatory observation, immersed in the society as completely as possible; structured interviews; systematic topographical surveys; creating a record of oral traditions. In 1935, he told Kroeber there were three questions that an anthropologist should always be asking herself: how does a culture become what it is? How does a culture or society determine the life of its people and, conversely, how do individuals influence their culture? And finally, are recognisable tendencies to change present? He warned students in the field never to look for an individual or psychological explanation for something they observed unless they had made every effort to find a cultural one.

Boas identified three characteristic mental functions present in all human groups, anywhere in the world: abstraction, inhibition and choice. 'All cultures, by virtue of being human, evince functionally equivalent capacities that are manifested differently due to environment and cultural context; as a corollary, variations in human biology do not constrain this cultural potential because it operates at the species level.'

His philosophy was shaped by critiques of social evolution and the notion of progress. 'The very concept of progress presupposes a standard toward which culture advances,' he wrote, adding that it is inescapable that that standard should be one's own. 'It is clear that this is an arbitrary standard and it is perhaps the greatest value of anthropology that it makes us acquainted with a great variety of such standards.' Cultural relativism, in the sense of resisting ethnocentrism or withholding judgement by any external standard, was a fundamental premise of his method, a methodological axiom, central to fieldwork, which became in his writing a polemical weapon.

He seemed to relish the iconoclasm that was companion to his cultural relativism. Refuting race as a cultural signifier, separating 'biological and cultural heredity', relabelling 'racial' traits as universal traits, was a major democratising, disruptive force. Once he had dismantled the unifying structure of social evolutionism, he was faced with cultural plurality, differences rather than similarities – cultures, rather than culture. 'Man is one,' as one of his students, Alexander Goldenweiser, observed, 'but cultures are many.'

But while, according to another student, Walter Goldschmidt, years after Boas's death, his liberalism was so stamped on anthropology that it became 'a natural characteristic of the field', that very liberalism could be a limitation because it was 'as if his belief that all people were equal rendered them all the same'. For Goldschmidt, it gave Boas's work an ominous sense of detachment. Despite his private experience of a devoted marriage, for example, he saw the institution in anthropological terms as primarily an economic transaction.

Facts were all. Loath to force data into a predesigned theory, Boas refused to synthesise them into analysis of any kind. He described ethnographic details and statistics with methodological rigour but until he knew everything, which could never happen, he wouldn't attempt anything more. He claimed it as a virtue – 'There are two kinds of people: those who have to have general conceptions into which to fit the facts; those who find the facts sufficient. I belong to the latter category' – but not everyone agreed. Radcliffe-Brown asked him to derive one generalisation from his long career; Boas would venture merely that 'People don't use things they haven't got.' That irrepressible critic Robert Lowie summed it up: 'I never saw a big man with fewer illuminating insights into things.'

Lowie's failure to persuade Boas to write or use a textbook, 'that solid basis of elementary fact, without which all one's ideas must be awry', created frustration that smouldered into career-long resentment. Until the late 1930s, Boas preferred to remain the oracle of anthropological experience. He published *General Anthropology* at last in 1938 (two years after he officially retired) and, while it provides a snapshot of what he saw as the purpose of anthropology, it is also an unconscious revelation of American society in the 1930s. He uses the phrase 'It is obvious that …' surprisingly often and refers in passing to 'the static picture of our time,', as if Western culture, unlike the primitive cultures his book was examining, was and would be permanent and unchanging. As in Malinowski's work, homosexuality does not merit a mention, despite long comparisons of various types of marriage, promiscuity and polyandry. Violence and rape are similarly excluded and war is explained in terms of status, not brutality.

He concludes with a passage that while outlining his central theory also exposes the maddening quality of much anthropological writing: 'The less pronounced the leading ideas of a simple culture, or the more varying the ideas of a tribe divided into social strata, the more difficult it is to draw a valid picture that does not contain contradictions. We cannot hope to do more than to elucidate the leading ideas, remembering clearly the limitations of their validity.

'The socio-political study, more than any other aspect of anthropological investigation, requires that freedom from cultural prejudice which in itself can be attained only by the intensive study of foreign cultures of fundamentally distinctive types that make clear to us which among our own concepts are determined

The image of a brilliant, maverick intellectual. None of Boas's evangelistic passion was diminished by age.

by our modern culture and which may be generally valid, because based on human nature.'

If Boas was an evangelist of cultural determinism who saw educating the next generation of anthropologists as one aspect of his role at Columbia, the other was being a public intellectual, guiding the American people through the complex ethical landscape of the twentieth century: 'Knowledge of the life processes and behaviour of man under conditions of life fundamentally different from our own can help us to obtain a freer view of our own lives and of our life problems.' Like his students, for whom even a week's research in the field could result in a monograph, he published prodigiously; his articles and letters regularly appeared in *The Dial*, *The Nation*, *The New York Times*, the New York *Evening Post*, *The American Mercury* and *The Atlantic*.

In *Anthropology and Modern Life* (1928), Boas addressed some of the issues confronting contemporary Americans, suggesting how anthropology might help answer them. Some of his conclusions have become givens of modern liberal thought, while others would be unfashionable today. He was concerned about changing relations between the sexes as newly independent young women increasingly prioritised careers over motherhood, but he believed education would be the key to positive social change. Eugenics, he argued, was a 'dangerous sword'; in an era before sophisticated genetics, there was no way of proving which traits other than hair and eye colour were hereditary and, anyway, how would you choose them? Not only that, as long as groups were divided by race, societies would be plagued by 'the desire for racial purity' already gaining ground in Germany. It was a sad truth that all societies seemed to possess the 'feeling of antagonism against other parallel groups'. Some people believed criminal behaviour could be predicted by analysing brain size, a spin on the nineteenth-century interest in phrenology, but Boas refuted this, arguing that crimes differed across societies and were nothing to do with biologically inherited characteristics. Indeed, he wondered whether any human traits at all were 'organically determined' or inherited; perhaps all were acquired culturally.

As the 1930s wore on, and the miasma of Hitler's Nazism permeated Europe, Boas commented with increasing fervour on patriotism and nationalism, aggression and peace. From World War I onwards he had feared what he called the 'terrible dream' of war, which promised glory but delivered division, hatred and misery. 'That one cherishes one's own way of life is a natural thing,' he wrote to his son in 1914. 'But does one need to nourish the thought that it is the best of all, that everything which is different is not good but useless, that it is right to despise the people of other nations?' In the 1910s, though, as an immigrant German Jew, he spoke out quietly, fearing jeopardising his position in the United States – and during the war and just after he was considered suspect even by some peers, forced to resign from the National Research Council in 1919 and expelled from the American Anthropological Association (which he had founded in 1902) the same year.

Having reasserted his control over 'his' discipline, he could afford in the 1930s to be more forthright about his conviction that what he called 'Nordic prejudice' and this 'shallow twaddle about race' would be a devastating global force. As well as writing, he joined anti-Nazi associations and, from 1933, helped bring Jewish scholars blacklisted by the Nazis to the United States and to establish themselves there. Ardent eugenicists, the Nazis rescinded his doctorate and banned publication of his works in Germany, condemning 'Jewish Science' – Freudian psychoanalysis and Einsteinian physics as well as Boasian anthropology. Boas responded in a statement signed by 8,000 scientists declaring that race and religion were irrelevant to science.

While his teaching commitments, administrative burden (he received 25,000 letters a year) and public role inevitably kept him from his own fieldwork, when time permitted, until he was in his mid sixties, he continued his research with the Kwakwaka'wakw people of British Columbia, with particular reference to their myths. 'One thing is unfortunate,' he observed of the tales told him by the women of the Comox tribe, in this area at least unable to shake off his ethnocentrism, they 'are so very coarse, to put it mildly, that three quarters of them cannot be retold'. Despite his own strictures on living immersed with the people and speaking their language, he continued to use interpreters and usually stayed in a hotel; but he went barefoot, wore a blanket he'd woven himself, and was an enthusiastic host of potlatches, or feasts. The Kwakwaka'wakw viewed him fondly, calling him Mullmumla-eelatre, the name of a large local rock, meaning 'if you put water on him the south-east wind will blow'. When he died, they sang mourning songs for him.

Boas retired in 1936 and became 'emeritus' in 1938, but retained his office at Columbia and his passionate interest in anthropology – and anthropologists – until his death in 1942. During his forty-year career as teacher and mentor, he sent students all over the world to study 'savage' and 'exotic' peoples and bring back the message that 'the mental processes of man are the same everywhere, regardless of race and culture, and regardless of the apparent absurdity of beliefs and customs'.

CHAPTER SEVEN

The Maiden

Ruth Benedict in the American Southwest, 1920s

———•———

Ruth Benedict, who studied anthropology under Franz Boas in the early 1920s, was not a natural fieldworker. Childhood measles had left her with impaired hearing that made learning and speaking other languages difficult, and she was painfully shy. Fieldwork was a challenge she had to overcome to become an anthropologist and a professor. The puzzle of academic work, intricate and painstaking, was what engrossed her: 'piecing together bits, filling in lacunae, and discovering correspondences and congruencies in the scattered and uneven accounts of vanished and of almost extinct cultures'. Despite this incompatibility with the field, she was sometimes overwhelmed by her experiences in the red and gold immensity of the American Southwest. In 1925, she wrote to a friend, 'Yesterday, we went up under the sacred *mesa* [flat-topped hill] along stunning trails where the great wall towers above you always in new magnificence ... When I'm God I'm going to build my city there.'

Once Boas was established at Columbia and began sending students out into the field, trainee anthropologists descended upon the American Southwest. The landscape was beautiful, the sun shone for much of the year, and from the 1910s onwards, Columbia students could board a train at Grand Central Station and find themselves, after a couple of nights in the comfort of a Pullman car, in Chicago or Santa Fe. Many were captivated by the aesthetics of Native American culture as well as the sense that it contained mysterious lessons that might mend the tainted post-war civilisation of the West. These young anthropologists saw primitivism as purity and authenticity. It was also considered an area in which white women travelling alone would be safe.*

White women had been researching the culture of Native American peoples for some decades by the time Ruth Benedict arrived. Matilda Coxe, who married the geologist James Stevenson in 1872, was one of the earliest female ethnographers. For sixteen years she accompanied and assisted her husband on his expeditions to survey Idaho, Colorado, Wyoming and Utah while both developed an interest in Native American culture; by 1879, Matilda had been appointed Volunteer Coadjutor in Ethnology by the Bureau of American Ethnology. From Oxford in 1884, that towering figure of nineteenth-century English ethnology, Sir Edward Burnett Tylor, thoroughly approved: 'If

---

* Until 1931, when a young doctoral student from Columbia, Henrietta Schmerler, was murdered by an Apache informant in Arizona.

his wife sympathises with his work, and is able to do it, really half the work of investigation seems to fall to her, so much is to be learned from the women of the tribe which the men will not readily disclose.' Matilda Stevenson was one of the founding members of the Women's Anthropological Society in 1885 and its first president; the society was disbanded in 1899, when women were admitted into the main American Anthropological Association, revived by Boas. When her husband died in 1888, John Wesley Powell made her the first paid female appointment at the BAE and she based herself near the pueblo of San Ildefonso in New Mexico.

The next generation were financially independent enthusiasts like Zelia Nuttall and Elsie Clews Parsons, dedicated to their work but, because of their independence, viewed with slight suspicion by their male colleagues. Nuttall, the inspiration for the flamboyant Mrs Norris in D. H. Lawrence's 1922 novel *The Plumed Serpent*, had become friends with Franz Boas when they both worked on the Chicago World's Fair in 1893. Born in San Francisco in 1857, with a self-made Irish father and a Mexican American mother, she briefly married a French anthropologist, Alphonse Pinart, in 1880. This was the period during which she first visited Mexico, working as Honorary Professor of Archaeology at the National Museum of Anthropology in Mexico City. Her work on the terracotta heads of Teotihuacan established her scholarly reputation, though she still had to contend with rivals – male – who tried to pass her work off as theirs; she became celebrated for finding important lost or misattributed manuscripts. In 1902, she moved permanently to Mexico City, where her home became a salon for scientists and intellectuals.

Elsie Clews Parsons, the daughter of a New York banker, was one of the first students at Barnard College – an all-female university set up in 1889 in response to Columbia's refusal to admit women – and became a committed feminist in her twenties. In books like *The Family*, *The Old-Fashioned Woman*, *Fear and Conventionality* and *Social Freedom*, published between 1906 and 1915 (two under a pseudonym so as not to upset the political career of her then husband, the Republican congressman Herbert Parsons), she outlined a radical agenda advocating trial marriages, divorce by mutual consent, and access to birth control.

Although she received her doctorate in education (alongside sociology, philosophy and statistics) from Columbia before the turn of the century, not until about 1910 did she develop the passion for anthropology, especially folklore, that would dominate the rest of her life. Ethnology, she believed, 'opens your eyes to what is under your nose'. She would spend the next thirty years either in Arizona, New Mexico and Mexico studying the Pueblo Indians, or contributing to and presiding over various anthropological organisations, culminating in her election as the first female president of the American Anthropological Association in 1941, the year she died. As the Boasian scholar Marshall Hyatt observed, it is 'no exaggeration to say that over half of what we know today about Pueblo Indian religion, folklore, and art was either learned or paid for by Elsie Clews Parsons'.

Parsons was not an uncontroversial figure. She worried that her unorthodox feminism, which encompassed her dedication to work, made her seem 'manly' but it was more that her adventurous personal life, funded by a private income, contrasted with her public role as a social critic; some observers marvelled at her breathtaking ability to have and eat cake. She was also not entirely ethically scrupulous in her professional life, for example accepting a ceremonial role in the Hopi tribe to improve her access to information. In her work she used the facts that were helpful to her – Native American women's knowledge of natural birth control was harnessed to her campaign, alongside Margaret Sanger, to promote the broader use of contraception – while omitting those that felt unauthentic, like the advice given to victims of snake bite – 'Stand in the river; send for white doctor!' – which was in her field notes but never published.

Ruth Benedict came into anthropology through Parsons, with whom she studied ethnology at The New School (Parsons was one of its founders in 1919) before embarking on a doctorate under Boas at Columbia in 1921. She was then thirty-four, unhappily married and searching for some sort of work that would give her life meaning. Her early childhood had been marked by the death of her father and the loss of her mother to bereavement; her heartbroken mother made (as Benedict would later put it) a cult of her grief, pushing her two daughters away. Lonely and ashamed, Benedict had turned to writing and, as late as 1928, still dreamed of becoming a poet rather than a professor. She was not rich enough to be careless of convention, like Nuttall and Parsons, but in her twenties she studied 'the lives of restless and highly enslaved women' of the past, planning a biography of Mary Wollstonecraft in which she would show that 'restlessness and groping are inherent in the nature of women'. Feminism, she declared, 'does not live by its logic' but instead in 'the bright stinging realms of their [women's] dearest desires'.

Anthropology was the first thing, to paraphrase her friend Margaret Mead, that made sense to Ruth Benedict. Steered by Boas, following in the footsteps of Matilda Stevenson and Elsie Parsons, in 1922 she went to study the rituals and poetry of the Pueblo Indians, Native American people living in adobe pueblos, or villages, tucked high into the fortress-like cliffs and mesas of New Mexico.

Native Americans had for generations been struggling against a hostile US government for land and autonomy and against settlers who pretended respect for their traditions but would have preferred to see them as a profitable tourist attraction. They presented their own challenges to the crowds of anthropologists who arrived at their reservations, eager to discover all they could about indigenous culture before (in their parlance) it vanished completely. When Carl Jung visited the pueblos in the 1920s, he spoke to a tribal chief, Ochwiay Biano (Mountain Lake), who

'Beautiful walled palace... the one homosexual thirst there is no getting by,' wrote Margaret Mead of her colleague, Ruth Benedict.

commented to him on how 'cruel' white people looked: 'Their lips are thin, their noses sharp, their faces furrowed and distorted by folds. Their eyes have a staring expression; they are always seeking something. What are they seeking? The whites always want something; they are always uneasy and restless. We do not know what they want. We do not understand them. We think that they are mad.' European people thought, he added, not with their hearts but with their heads.

Understandably, Native Americans could be suspicious and reluctant informants, fearing curses, recrimination and loss of status if they collaborated with these uninvited, nosy and often unscrupulous guests. Ethnologists might pay for information with money* or tobacco – Matilda Stevenson notoriously used whiskey – promising confidentiality to those who did cooperate. Parsons thought nothing of evading the Pueblo 'code of secrecy' by taking notes in a station with 'a Santo Domingo man who succeeded in eluding his pottery-selling colleagues between trains'.

Robert Lowie was one of Boas's first students to do his graduate research in the field and his experiences in Idaho in 1906 and Alberta in 1907 reveal some of the difficulties early fieldworkers faced. He made the mistake of trying to rush into interaction when he arrived at a Shoshone reserve and saw a group gathered to collect their government rations; they either ignored him or laughed at him as, increasingly desperately, he promised larger and larger amounts of the tobacco he'd brought to trade for information. Eventually someone took pity on him and he was taken to meet an aged aunt who, the following day, gave him his first story in Shoshone. The next time, he walked through the camp making cat's cradles, hoping to entice tribespeople to interact with him. Thinking it strange behaviour for a white man, they sent someone to find out what he was doing and Lowie promptly hired him as an interpreter.

---

* Later, Audrey Richards would note her surprise at learning that American anthropologists paid their informers $4 an hour. British anthropologists did not buy information.

The travelling could be hard. It took Lowie longer to get from one reservation to another within the state of Idaho than it did to get to Idaho from New York, including over a hundred miles in a stage coach; on another occasion he travelled 252 miles in an open-oared scow and then a wood-burning steamboat. Living conditions were basic, with dogs and children coming in and out of the tepee at all hours and no privacy, ever. Bedbugs were a plague. Research pitfalls were everywhere. Lowie recalled a young anthropologist studying the Blackfoot tribe, faithfully recording each step of the process as a squaw put up a tepee. But the student later learned the woman he was watching was a Sioux married to a Blackfoot and her method was atypical. Lowie noted, 'the task is not merely to set down what we see or learn from interlocutors but also to define *what* it is that we have seen and heard'.

The objects of anthropological research had their own methods of reprisal. Soon after Matilda Stevenson's death in 1915, a young archaeologist near her ranch at San Ildefonso told a local indigenous man the name she said the Indians had given her, meaning Little Flower. The man laughed and replied, 'That doesn't mean "Little Flower", it means "Big Bottom"!'

Benedict acknowledged that her job required 'a certain tough mindedness and a certain generosity' as she sought to bring a new morality to anthropology, respecting the people who were the focus of her research. As a woman, she understood the constraints and coercions societies imposed on their weaker members; her own experience of marginalisation gave her a greater insight into

– and impetus to defend – other marginalised groups. But even she sometimes needed advice. Frustrated by her inability to learn more about faith and ritual, she appealed to Jaime de Angulo, the musician, linguist and ethnologist, for help in finding 'an informant who would be willing to give tales and ceremonials'. Angulo was appalled at the idea of persuading someone to give up the deepest secrets of his people. 'God! Ruth, you have no idea how much that has hurt me,' he replied. 'Do you realise that it is just the sort of thing that kills the Indians? I mean it seriously. It kills them spiritually first, and as in their life the spiritual and the physical element are much more interdependent than in our own stage of culture, they soon die of it physically. They just lie down and die. That's what you anthropologists with your infernal curiosity and your thirst for scientific data bring about.'

It was not until the late 1920s, when she began to compare the Pueblo Indians with the Plains Indians, that Benedict started making her most celebrated observations on Native American culture. Generally speaking, the Pueblo tribes, among whom are the Hopi and the Zuñi, are peace-loving, village-living agriculturalists inhabiting the cliffs of the deep Southwest; the Plains people, including the Apache, the Comanche and the Crow, are more nomadic, warlike bison-hunting tribes from the central American prairies. Benedict identified the Pueblo Indians, who were matrilineal, egalitarian, tolerant and cooperative, as Apollonian, while she saw the Plains Indians as Dionysian, male-dominated, aggressive and excessive.

For the most part she left her readers to draw their own conclusions but, for her, the Pueblo people provided an alternative model to the homogeneity and hypocrisy of contemporary American society. She also admired the Blackfoot, a Plains tribe with whom she worked in 1938, who were individualistic but had a working system of social welfare – something Benedict, as a New Deal liberal, was pleased to see functioning successfully. It was what she and her colleagues called a 'high-synergy' society, which successfully satisfied the needs of both the group and its individual members. Yet even while advocating cultural relativity, she admitted its limits. She described how Zuñi women who were unfaithful were not punished by their husbands, while among the Apache the wronged man bit off the end of his wife's nose. From an anthropological point of view, she pointed out, both patterns 'functioned' successfully. One response was 'Apollonian', the other 'Dionysian'. But it was clear which culture she approved of.

Robert Lowie explored this topic in an essay on ethnographic empathy in which (although written after her death) Benedict seems present as an invisible reader. The corollary of the idea that one must not impose moral evaluations on the subjects of an enquiry is that one must not impose moral evaluations on anyone: 'If the Crow worship of a medicine bundle [a wrapped bundle of sacred items used in Native American rituals] can be understood, why should one condemn as the worship of idols the innumerable shrines that dot the Bavarian countryside?' Although the scientist must not pass moral judgement, context is all-important. Cannibalism, infanticide or the abandonment of aged parents can all be viewed differently according to the circumstances in which they occur: 'It is an ethnographic commonplace that the aborigines who are crudest in their general mode of life generally do not eat human flesh, whereas those who most greedily indulge in it are comparatively advanced.' Besides, he asked elsewhere, 'What would be thought of a modern zoologist who should denounce the wickedness of a rattlesnake? Surely an ethnologist may be excused from declaring that he does not favour indulgence in human flesh when he describes cannibalistic rites.'

Benedict described the vision quest, a ritual of adolescence in the Pueblo cultures leading to a young man being told his true name and meeting his guardian spirit but also something the mature man in the Plains tribes looked for perhaps before battle or during mourning. After a period of fasting, the quester sought out his vision with varying forms of violence and surrender: among the Blackfoot, the self-mutilation involved was known as 'feeding-the-sun-with-bits-of-your-body'. It often included the use of peyote or other stimulants and ended with sex. Among some Plains people, visions could be bought and sold. 'No man sees the logic of another's symbols,' Benedict concluded plangently. The Indian and the white man 'played, both of them, the identical game – the game of prestige. One played it with songs and visions and the giving of goods for counters; the other played it for land. And if the red man's counters were harmless and dispossessed no one of food or shelter, on the white man's counters have hung progress and the glories of civilisation.'

She was following her heart in returning to the concept of religion, rather out of academic fashion since the late nineteenth century. Decades earlier, the popular philosopher and polymath Herbert Spencer had proposed that faith and ritual sprang from ancestor worship; Edward Tylor at Oxford, developing the idea of animism, believed the concept of the soul originated in dreams and visions; James Frazer compared the myths of many cultures and proposed that faith in magic would end in faith in science; in Paris at the turn of the century, Émile Durkheim saw religion through a sociological lens, as the collective outpouring of excitement by a crowd. Benedict wondered if religious feeling was stimulated by fear, dissatisfaction or some innate sense of mysticism; she noted that it existed in every observed society. All cultures believed either that objects and creatures possessed supernatural powers or that the universe exerted a power over the things and people in it. Further she argued that myths were not just archaic stories, remnants of the distant past, but incorporate cultural details that revealed everyday contemporary life, changing to reflect present-day culture as well as its history.

Always she sought to look beneath the surface of her observations: 'As a matter of fact, agriculture and economic life in general usually sets itself other ends than the satisfaction of the food quest, marriage usually expresses other things more strikingly than sex preference, and mourning notoriously does not stress grief.' As she knew from her childhood, watching her devastated mother fail to come to terms with her father's death, mourning could have very little to do with grief; and as she knew from her marriage, which was only briefly loving, matrimony could have very little to do with desire.

One of the aspects of Native American culture that most fascinated the female anthropologists of this generation, especially Benedict, was its attitude to gender and sexuality. Many, like Elsie Clews Parsons, Ruth Underhill and Gladys Reichard, as well as Benedict, were either divorced or deliberately unmarried, seeking independence from traditional gender roles they found oppressive; others, including Benedict, who felt themselves to be 'misfits' or 'deviants' in their own society because of their 'sex preferences' were further consoled by discovering there were other ways of being. 'It is interesting to note that, in the languages of the world, gender is not by any means a fundamental category,' Boas had observed in his *Handbook to Indian Languages*. The Bantu languages of Africa use many distinct classificatory groups; the Iroquois distinguish between men and all other nouns; the Algonquin separate nouns by animate and inanimate. Benedict and her colleagues were reassured by the cultural differences anthropology revealed to them.

Matilda Stevenson, in her years of research with the Zuñi, developed a deep affection and admiration for a six-foot transvestite called Wé-Wha, through whom she introduced laundry to the tribe. Stevenson believed Wé-Wha was physically male but psychologically and socially female; along with the rest of the tribe, she referred to her using female pronouns. Wé-Wha was a man-woman, or *ta'mana* in Zuñi, usually but not always having a male sexual role and a female gender role. According to Elsie Parsons' research, there didn't seem to be any social penalties for being a man-woman or, as they were generally known, *berdache*: to the Indians, they were just another gender. Some were 'pretty', others not at all effeminate; two of the nine known to Parsons' informant might have been 'simple'. They were initiated with the other boys but could marry men and, when they died, they were buried in the men's side of the cemetery, wearing a woman's skirt over a man's trousers. Importantly, they had a special spiritual position in Native American societies. Zuñi men-women became bird gods, physical links between the feminine land and the masculine sky, the worlds of the living and of the spirits.

To these female anthropologists, the *ta'mana* was a new type of identity they hoped might inspire open-mindedness and empathy in contemporary American society. In a 1932 paper entitled 'Anthropology and the Abnormal', Benedict showed 'how certain personality traits considered abnormal in our culture may be highly valued in another culture and are therefore no longer considered abnormal'. Implicit in their celebration of the Native American tolerance for the man-woman, and of matrilineal societies, was a critique of the polarised, patriarchal society in which they themselves had been raised. Even when they studied patrilineal societies, their strong identification with the women – as the father figures of their culture, such as Boas and Tylor, advised – could become tacitly mutinous.

Many years later, Clifford Geertz would call the sense of purpose that animated Benedict 'edificatory ethnology'. She believed an anthropologist had a responsibility to use what she had learned in the field to promote a free and democratic society when she returned, broadening the perspective of her compatriots and provoking a critical reappraisal of their culture. The 'deviant's tragedy', as she put it, was to be born into an unsympathetic society but it helped to know that cultural traits weren't inevitable: at other times and in other places, people and institutions had been and would be more tolerant of difference and more humane in their treatment of it. This human plasticity was something she always stressed, alongside the diversity of culture.

Benedict's complicated private life profoundly influenced the path of her career. Boas, whom she called Papa Franz, became a beloved father figure to her, a personal as well as professional support and mentor. By the 1930s, he was increasingly frail, his recurrent health problems intensified by the tragedies of losing three of his six children and his beloved Marie, who was hit and killed by a car in 1929. 'Your physical weakness distresses me daily,' Benedict wrote to him in 1939. 'There has never been a time since I've known you that I have not thanked God all the time that you existed and that I knew you. I can't tell you what a place you fill in my heart. It seems paltry to send you good wishes but I send them constantly nonetheless.' Hortense Powdermaker was surprised to find that Benedict had photos of Boas all over her apartment, as well as in her office.

Benedict met Margaret Mead in 1922 when, as a reserved graduate student assisting Boas, she taught her first course at Columbia; their lives and work would be closely entwined for nearly thirty years. For the first years of the friendship they were lovers, too, although Benedict was also carrying on affairs with Tom Mount and another colleague, Ruth Bunzel. In 1931, she and Stanley Benedict separated after years of unhappiness and she began to view herself as a lesbian, an epiphany that released her from the anguish of feeling herself to be a deviant and set her free creatively. In the same year, she formally took up a teaching position at Columbia as an assistant professor, a role her husband had preferred her not to accept. (Her salary, at £3,600, was far less than that of a male visiting scholar.) In the early 1930s, she grew close to the young chemist Natalie Raymond; later she fell in love with the psychologist Ruth Valentine, with whom she would live for the rest of her life.

When Boas fell ill in the 1930s, Benedict informally chaired the anthropology department in his place but, although her friends hoped she would be chosen as the first female head of department in the country when he stepped down, she was passed over with the consolation of an associate professor role. Alfred Radcliffe-Brown, then at the University of Chicago, was apparently in the running for Boas's position but he was deeply unpopular at Columbia, not least with Benedict, who described his condescension and arrogance in letters to Mead. Not until 1948, when Boas's successor, Ralph Linton, left Columbia to take up a role at Yale, was Benedict granted tenure and made full professor. By then she had taught at Columbia for more than twenty-five years (unpaid for many) and been president of the American Anthropological Association. Even so, as a woman, she was not allowed to enter the university's faculty dining room.

As a teacher, she was full of quiet charisma, inspiring the devotion of her students and colleagues and exerting an almost spiritual influence over them: 'With her silvery aura of prestige, dignity and charm … [she was] like a symbolic representative of the humanistic values of the Renaissance,' wrote Victor Barnouw, one of her students. Her way of expressing herself could be gracefully oblique but she 'had a way of talking about "primitive" peoples as if she could see an X-ray of their souls projected upon an invisible screen in front of her'. Tall and slender, she looked like the 'Platonic ideal of a poetess', with the 'hooded grey eyes of an aristocratic eagle'.

According to Barnouw, Benedict herself was a combination of Apollonian and Dionysian: a poet and dreamer on one side, a scientist and scholar on the other. These two impulses were united in her masterpiece, *Patterns of Culture*, which came out in 1934 and had by 1973 sold 1.6 million copies in English. Margaret Mead called it 'an approach to an unfolding world': fresh, wise and compassionate. For the rather more jaded Alfred Kroeber, it was

Dignified, generous and serene, Benedict had 'the hooded grey eyes of an aristocratic eagle'.

'propaganda for the anthropological attitude'. 'No man ever looks at the world with pristine eyes,' wrote Benedict in its introduction. Western civilisation might dominate the world in which she lived but, she argued in austerely lyrical prose, anthropology and the cultural relativity that accompanied it would give people 'a more realistic social faith, accepting as grounds of hope and as new bases for tolerance the coexisting and equally valid patterns of life which mankind has created for itself from the raw materials of existence'.

Benedict's passionate engagement with her subject, derived from her early sense of alienation from the culture she inhabited, attracted criticism from some colleagues. Robert Lowie, who also studied the Plains Indians, thought she idealised the Zuñi; for him, their harmonious facade masked factionalism and rampant gossip. Others argued that the Plains Indians were not nearly as Apollonian as Benedict presented them to be. Barnouw was among those who critiqued her attachment to relativism, arguing that the tolerance for cultural variation she extolled would necessitate acceptance of racial segregation in the Jim Crow South and the wilder excesses of the Nazi regime.

Those anthropologists who believed the goal of anthropology was searching for objective laws of social behaviour were always going to find fault with Benedict's determination to combine scientific methods with the concerns of the humanities. Radcliffe-Brown accused her of having a 'rags and tatters' approach to culture, because she rejected his functionalist approach in favour of Boas's diffusionism.* Other students of Boas took literally his emphasis on facts – spending their time diligently gathering potsherds, blueberry pie recipes and string figures without the 'driving central purpose' that motivated their mentor – but

---

* For her part, she found Radcliffe-Brown maddening. 'If only he held to a high standard of achievement and required language control, intimacy with total culture, fundamental understanding of kinship, I could understand his scorn of work so far done in America. He could scorn work in broken cultures too. But to thrust this kind of work under my nose as the salvation of the world, it's sad,' she wrote to Mead. 'I don't think Brown is fighting for good work over against bad, but for work done by disciples over against work done by non-disciples. And that's fatal.'

Benedict had the courage to find and pursue meaning among the data. Although Barnouw accused her of projection, the unconscious transfer of one's desires and emotions onto another person or people, for him 'the aims she pursued and the kinds of questions she asked' made her great. Her 'dignity lay in the fact that she went after important issues'.

Benedict was aware of this criticism, speaking in 1946 and 1947 about her belief that although anthropology was a science, the humanities were anthropologists' greatest resource in the study of the mind of man: 'I have the faith of a scientist that behaviour, no matter how unfamiliar to us, is understandable if the problem is stated so that it can be answered by investigation and if it is then studied by technically suitable methods. And I have the faith of a humanist in the advantages of mutual understanding among men.'

One of her strengths as a writer was juxtaposing 'the all-too familiar and the wildly exotic in such a way that they change places', according to Clifford Geertz. He described the way she drew attention to morbid practices like cannibalism, scalping and megalomaniac potlatch (competitive feasting in which everything is given away, especially among the native American tribes of the north-west of America) in order to make subtle comparisons with twentieth-century Western culture, marked as it was by economic devastation, political instability and violence.

Edward Sapir, who specialised in linguistics, was a close friend of Benedict's in the 1920s, when both were hopefully submitting their poetry to the same reviews and magazines. He argued that societies were only 'genuine' when their dominant drives and cultural traits were in harmony. Benedict, more sensitive to the darker side of human nature, suggested that 'cultures may be built solidly and harmoniously upon fantasies, fear-constructs, or inferiority complexes'.

When the Second World War began, full of fantasies, fear-constructs and inferiority complexes, Benedict strapped on her metaphorical armour. For many years she, like Boas, had been countering prejudice in American society. In her 1942 book *Race and Racism*, she quoted the biologist Julian Huxley (brother of Aldous), pointing out that distinctions in skull or nose size, eye colour, even blood type were greater within groups than between them, and emphasising the effect of long-past migration patterns: 'multiple ancestry is at least as important as common ancestry in considering the nature and origins of any group'. She argued succinctly that 'race is a matter for careful scientific study; racism is an unproved assumption of the biological and perpetual superiority of one human group over another'. In 1943, she and a colleague at Columbia, Gene (Regina) Weltfish, produced a pamphlet entitled 'The Races of Mankind', which outlined, in the simplest of terms, the scientific case against racism. It was intended for American troops but, following a Congressional outcry against its 'communistic' tone, never circulated.

'We in America are willing to pay enormous prices lest liberty be lost on our continent,' Benedict wrote elsewhere, during the war. 'The only argument is how best to keep ourselves strong and uncontaminated. For this great end we must be clear in our minds that the way to keep ourselves from the taint of our enemies is through the defense of civil liberties. We must be sure that we do not curtail them in the fields already allowed, and we must extend them to other fields not now recognised. For liberty is the one thing no man can have unless he grants it to others.'

At the behest of the US government, alongside Robert Lowie, who wrote about Germany, Benedict turned her anthropological gaze on Japan, using the techniques previously focused on 'primitive' people to evaluate the 'learned cultural behaviour' of a sophisticated totalitarian nation. She hoped to discover what was needed to create 'a world organised for peace'. *The Chrysanthemum and the Sword* (1946) was, by necessity, written without visiting Japan but provided important insights into Japanese culture to American soldiers and diplomats, for example about the role of the Emperor and the distinction between guilt (a dominant force in Western and Christian cultures) and shame (in Asian). It would become a bestseller all over the world, especially in Japan.

The people who knew Ruth Benedict best were deeply touched by her gentleness and generosity. 'In her later years, beset by phone calls, interruptions, manuscripts to read and refugees to interview, she moved through her book-lined office with a kind of sad, blurry nobility, overworked but always with a renewing buoyancy of spirit, kindliness and a deep sense of social responsibility.' Her work and personality told 'the story of a woman's struggle to find a place for herself and to make a meaningful life for herself in a society in which she felt like a misfit'. Through anthropology this, triumphantly, Benedict was able to do.

CHAPTER EIGHT

The Child

Margaret Mead in Samoa, 1925

——•——

'In all my years of fieldwork, each place where I have lived has become home,' Margaret Mead observed in her memoirs, *Blackberry Winter* (1972), but her most permanent adult home in the end would be her office, tucked under the eaves of the American Museum of Natural History alongside the sealed and reinforced units storing marvellous objects from all corners of the world. Mead got her first job as assistant curator at the museum in 1926 and remained there more than forty years, inhabiting a little attic room looking out over the red-tiled rooftops of the Upper West Side. Cotton curtains hung at the windows, there were Samoan mats on the floor, piles of notebooks, files, manuscripts and mimeographed love letters everywhere, rolls of film, feathered headdresses, her trusty typewriter, an ex-husband's old revolver, and a large map of the world pinned to the wall, for planning her next expedition.

In 1969, when she retired from the museum, Mead was sixty-eight and the only anthropologist most Americans would

"The one I liked best is Margaret Mead."

This 1958 cartoon by Ed Fisher for the Saturday Review *pokes gentle fun at the ubiquity of America's best-known anthropologist.*

have recognised if they saw her in the street. Short and stocky, with cropped grey hair and glasses, she favoured a felt cape and sensible shoes and used a forked walking stick acquired on her travels – equally ready to pin down a snake as pick up a 'native' child and give her a hug. (Her cape and stick now stand on permanent display in the museum.) She saw herself not just as an anthropologist working in the field but, like some kind of Prometheus, bringing her knowledge back as a mirror to hold up to the people she had left behind. Her mission over her long career was to study 'the lives of other people, faraway peoples, so that Americans may better understand themselves'.

That mission began in 1923, when Mead walked into her first graduate class in anthropology at Columbia, where a reserved Ruth Benedict was assigned as her teaching assistant. Mead described herself at twenty-two as disarmingly innocent but she was familiar with the academic world because her father was a professor of finance and her mother a sociologist working with immigrant families; both had studied under the influential sociologist and economist Theodore Veblen. Franz Boas was by then in his sixties and Mead was part of the 'grandchild' generation of his students. The first generation, Robert Lowie, Alfred Kroeber, Edward Sapir and others, described the disciplinarian Boas assigning them homework in a language they couldn't read; by the time Mead arrived, he had mellowed enough to let her choose where to do her fieldwork – Polynesia – but still dictated that her study would be adolescent girls.

There were no guidelines for students setting off into the field. 'No one considered whether we could stand loneliness,' wrote Mead years later. 'No one inquired how we would get along with the colonial or military or Indian Service officials through whom we would have to work; and no one offered us any advice.' Advice, in fact, was considered almost cheating. When a more experienced Mead gave a female student headed for Africa 'some basic instructions on how to cope with the drinking habits of British officials, anthropologists in London sneered'. There was a logic: if unprepared young fieldworkers 'do not give up in despair, go mad, ruin their health, or die, they do, after a fashion, become anthropologists'. The only thing Boas told her was to 'be willing to seem to waste time just sitting about and listening'.

Mead was newly married but that was the extent of her life experience. She had never been abroad, never stayed in a hotel alone, spoke no language other than English fluently. She did not know how to swim. She was young enough to mind that because she had been told that silk rotted in the tropics, she took just cotton dresses 'including two very pretty ones', only to find all the 'navy wives' she encountered in silk. In her luggage, alongside the dresses, were a small strongbox for her money and papers, a portable typewriter, a Kodak camera, a torch and piles of paper and pencils.

In August 1925, she travelled with Benedict across country from New York, bidding her farewell at the Grand Canyon; Benedict was heading on to Zuñi. Bonded by their studies and their devotion to Papa Franz, the two women had become intimate friends, despite the fourteen-year age gap between them. When Mead had had trouble finding funding for her doctorate, Benedict gave her $300 towards it, calling it a 'No Red Tape Fellowship'. Mead – only twenty-four, but already married for two years to her childhood sweetheart, Luther Cressman, a theology student who would later become an anthropologist – was being pursued by the linguist Edward Sapir, another of Boas's disciples and Benedict's colleague. Recently widowed,

Sapir had been trying to persuade Mead she would 'do better to stay at home and have children [with him] than go off to the South Seas', but their relationship was nearing its end. On the train journey, as Mead remembered it, she and Benedict discovered they 'preferred each other'.

After Benedict disembarked, Mead continued on alone, writing passionate letters to her in which she at once agonised between the competing demands of her husband and Sapir and exulted in the secret love she shared with Benedict. 'Every memory of your face, every cadence of your voice is joy whereon I shall feed hungrily in these coming months,' she wrote when they parted and, throughout her months in the Pacific, in her correspondence with Benedict, their love and their joint passion for their work sustained her. 'Risk my love – sweetheart, sweetheart, what nonsense you do talk – and will the birds forget to come north in the spring to the land of their desire?' she asked six months later. 'When I do good work it is always always for you – That's my wishing … I've a hundred details I should be writing about, but if I were there I'd kick all the mss. and proofs under the table and bury my face in your breast.'

In August 1925, Mead set sail from San Francisco to Hawaii, where a family friend evidently believed she would be a sort of Sunday School missionary, sending her off with a hundred squares of torn white muslin 'to wipe the children's noses'. She was almost surprised to find no one waiting on the dock to greet her when she arrived in Pago Pago, the bustling capital of American Samoa and coaling centre and repair station for the US fleet in Asia. After six weeks living in a hotel learning Samoan, she embarked for the island of Tau, also under American administration, where she settled into the porch of the naval dispensary, a perfect location for anthropological research since all the children came and went freely there.

In letters home, Mead described her favourite time of day, when with a group of fifteen or so girls and children she would walk through the village to watch the waves at sunset. They would go back with her to her room and keep her company in the evening, chatting incessantly. 'They dance for me a great deal,' she reported, 'by no means in a puritan fashion.' Like Malinowski in the Trobriands, she travelled round the island 'like a visiting young village princess', summoning 'informants to teach me anything I wanted to know; as a return courtesy, I danced every night'. The idea clearly pleased her. 'Your sister is high priestess of a Polynesian village,' she wrote to her sister, only half joking. 'Grand thought.'

Her correspondence was her only relief from terrible loneliness, missing (as she put it) not sex but intimacy and gentleness. When a letter arrived from Sapir informing her he'd fallen in love with someone else, she gathered up all his letters and made a bonfire of them on the beach. Benedict, herself vulnerable to depression, urged Mead to 'develop all the expedients you can against weeping – companionship is only one of them'. Unlike Mead, she didn't favour giving young fieldworkers too much practical advice, but she did have methods for staving off the 'blue devils' in the field, ranging from brushing her teeth and gargling 'with every onset, to playing you're your own daughter for a year' – although hard work, she'd found, was the most effective relief.

Contributing to a 1931 anthology for girls called *All True! The Record of Actual Adventures that have Happened to 10 Women of Today*, Mead remembered how unsure she was on her arrival in Tau of succeeding 'in this strange kind of adventure, this adventure of shedding all one's own ways of eating, sleeping, talking, laughing, just as if they were an old skirt instead of the most important part of me, and putting on the attitudes of a Samoan girl, as easily as if they were only a party dress'. She seems to have felt that to endear herself to her hosts she should become as much like them as she could, that she would earn their trust by apparently adopting their culture, and so she put on fringed Samoan bark dresses and hibiscus garlands, had her hair rubbed with scented oil, and danced and feasted with her new friends. Her tactic was so effective that she received several proposals of marriage; when she turned down one suitor, a visiting chief, he observed regretfully, 'White women have such nice fat legs.'

What most worried her was that she had no idea whether she was doing her work right. 'I've got lots of nice significant facts,' she told Benedict, but she doubted they amounted to anything.

---

* Robert Lowie said that the greatest compliment of his life was one he had overheard one Native American say to another: 'You see that white man over there? He looks like any other white man, but when he comes to the camp fire, you'd never know him from an Indian.'

Mead wearing a bark dress with Samoan friends in 1925. Her hosts treated her 'like a visiting young village princess'.

'I'm going to get a nice job giving change in the subway when I get home.' She wrote to Boas just before she arrived on Tau, outlining her research plans, and received his 'reassuring' reply just as she was preparing to leave. His foreword to the book she wrote about her time in the South Seas, *Coming of Age in Samoa*, reveals a little more about what he hoped she might learn when he sent her off into the field: 'Courtesy, modesty, good manners, conformity to definite ethical standards are universal, but what constitutes courtesy, modesty, good manners and ethical standards is not universal.' Mead's study of adolescence in Samoa, Boas wrote, confirms 'the suspicion long held by anthropologists, that much of what we ascribe to human nature is no more than a reaction to the restraints put upon us by our civilisation'.

Using the 'laboratory conditions' (her words) provided by a 'primitive' society that lacked the 'civilised' complexities of written language and history and organised religion, in Samoa Mead sought to conduct a controlled experiment that would provide insight into 'the difficulties of childhood and adolescence' (Boas's words) of 1920s America. Like Malinowski, she granted herself a fundamental role as the interpreter, almost the discoverer, of the culture she had observed, marvelling in 1952 at the 'extraordinary historical accident' – her arrival in Samoa – that gave her Samoan companions an existence in books they would never read, 'far beyond the world that their imaginations could have dreamed of'.

What Mead discovered, through frank interviews with about fifty girls just younger than herself, in three villages, was that their culture permitted Samoan teenagers sexual freedom before marriage. Her new friends deferred 'marriage through as many years of casual love-making as possible'. By demonstrating that adolescence didn't cause anxiety in Samoa – growing up there, she wrote, was 'so easy, so simple a matter' – she believed she had shown that American culture made its teenagers neurotic and 'maladjusted'. Her conclusions were that sex roles were determined by culture – nurture, not nature – and that freedom from sexual guilt led to happier, better-adjusted adults. 'Civilisation' brought with it hypocrisy and contradiction, the 'intensity' of the nuclear family as opposed to the loving, tolerant generality of the Samoan village, segregation between the sexes and taboos around discussing sex, menstruation, birth and death.

Given her inexperience, it is hard not to agree with at least some points made by her most serious critic, Derek Freeman, who in 1983 (five years after Mead's death) fundamentally challenged her Samoan findings.* He argued that she had arrived with a preconceived opinion (that premarital promiscuity was widely accepted in Samoan society), was slapdash and ill-disciplined in her research, stayed for a shorter time than planned because she was lonely and used her interviews less as empirical investigation than to confirm what she already thought she knew. She skimmed over her subjects' Christianity (missionaries had arrived in Samoa long before the anthropologists) because it would read less well to an audience eager for exoticism and her eagerness to please Boas shaped her findings. Freeman quoted the letter Mead wrote to Boas telling him she planned to 'write conclusions and use my cases as illustrative material'; this was the letter he had answered with reassurance.

Freeman, who had taught in Samoa in the 1940s and returned again as an anthropologist in the 1960s, believed that

---

* He was not the only person who saw Mead as unreliable. In private correspondence, Hortense Powdermaker commented that she found her 'fundamentally dishonest'.

Mead had been the victim of an elaborate hoax: that the girls she had befriended, realising she wanted to hear tales of sexual freedom, had 'just fibbed and fibbed to her' about spending nights with boys, never dreaming she would take them seriously. After more than a century of lustful tourists, they would have known Westerners were inordinately interested in their sexual customs. He suggested she had not properly understood that her closest friend in Samoa, Fa'apua'a, was a *taupou*, a ceremonial role for a young woman of high birth required to remain a virgin until marriage, and that she didn't recognise Samoan people joking when asked about sex, as they did in many cultures.

Of course, Freeman had his own agenda – he believed that Samoans, far from being laid-back with no interest in status or success, were actually competitive and that Mead, by focusing on children and teenagers, had misunderstood the power structures at play. Furthermore, he was debating events that had taken place more than forty years earlier, when values and customs were different; his primary informant, Fa'apua'a, had become a zealous Christian with an interest in countering Mead's portrait of a sexually uninhibited Samoan society and her role in it; and, even if he had wanted to talk to young girls about this, he would not have had Mead's access to them. In her defence, far from viewing Samoa as a prelapsarian paradise, Mead found it rather boring, limited by its indolence and lacking the creativity of more dynamic cultures like her own.

But all this criticism was far off when Mead returned from Samoa (having fallen in love on her return voyage with Reo Fortune, a New Zealander who would become her second husband), wrote her book and became a celebrity overnight. *Coming of Age in Samoa: A Psychological Study of Primitive Youth for Western Civilisation*, with its credibility-enhancing foreword by Boas and a puff, arranged by her publisher, from Bronislaw Malinowski, was published in 1928 with an illustration of a bare-breasted girl and boy on the dust jacket. It would be the best-selling anthropological book in the world almost until she retired fifty years later.*

The book came out into an America obsessed by youth, simultaneously perplexed by and slightly disapproving of the changing morals and mores of the younger generation, but eager to partake of them. The Jazz Age and Prohibition were in full swing. In movies, bob-haired flappers tucked flasks of bootleg gin into their garters and necked uninhibitedly with their boyfriends; Margaret Sanger was campaigning to make contraception accessible and acceptable; young women for the first time saw a job and independence as a viable alternative to marriage and a family. Mead's study, with its scientific gloss and tone of utter conviction alongside its evocative portrait of a tropical idyll,

---

* In the foreword to the 1973 edition, Mead observed that two things about her first book merited comment nearly fifty years later: her assumption that Samoan life as she witnessed it would quickly and inevitably disappear, and her failure to address her book to Samoan readers who might, she now realised, feel excluded by a book written about, but not for, them.

showed Americans eager for change that another kind of society was possible, even desirable.

Not everyone viewed these developments as positive. Some anthropologists – even some Boasians – were concerned by the impatience with which American women were apparently jettisoning their traditional roles as wife and mother. In turn, Mead and Ruth Benedict found their older colleagues, Alfred Kroeber and Robert Lowie, masculinist and patronising. Edward Sapir, whose conservative views had alienated him from Benedict, once a close friend, and his former lover Mead, criticised Mead's inaccurate fieldwork and called the book 'cheap and dull'. In a 1928 essay entitled 'Observations on the Sex Problem in America', he also derided the new freedoms that he thought unnaturally separated sex from love and, in a not-so-veiled dig at Mead, denounced the idea of a therapeutic quest for personally enriching sexual experiences. As a spurned lover rather than an academic critic, he was vituperative in private, telling Benedict that Mead was 'a loathsome bitch ... a symbol of nearly everything that I detest most in contemporary American culture'.

In the end, Malinowski, whose *Sex and Repression in Savage Society* had come out the previous year (which perhaps was why he had been willing to praise *Coming of Age*), was also dismissive of Mead's methods, earning her lasting rancour, although she did acknowledge he had played a part in making anthropology accessible to a wider public; slightly disingenuously, given her own adventurous emotional life, she was privately critical of his womanising, too. Malinowski and other British anthropologists dismissed her writing as novelistic – the most damning of judgements in the academic world.

Mead was undaunted. Although the wariness (and perhaps envy) of her peers prevented her from achieving any distinguished academic role, she remained curator at the Museum of Natural History, continuing with her fieldwork and returning after each episode to write up what she had observed in her attic office and enlighten her fellow Americans. Despite the disdain of her tenured colleagues, her books sold better than any of theirs. Like her contemporary, Amelia Earhart, she became a role model and inspiration for modern girls who dreamed of adventure rather than domesticity.

In 1928, Mead and her second husband, Reo Fortune, a student of Malinowski, travelled to Manu in New Guinea. The eight months they spent there, which resulted in her 1930 *Growing Up in New Guinea* (and Fortune's 1935 *Manus Religion, An ethnological study of the Manus natives of the Admiralty Islands*), was 'the best field trip we ever had'.* Her focus on families, maternity and child-rearing, especially making parents aware of their children's need for love and security, was something that would mark the

---

* Beatrice Blackwood, an English anthropologist who met Mead in Sydney immediately after it in 1929, 'disliked the woman intensely … For one thing – a person who spends six months in a place (during one month of which I afterwards discovered she lived with a white woman nursing a sprained ankle) – and then says she speaks the language perfectly and knows all about the natives – always makes my hair stand on end.' Larson, *Undreamed Shores*, 216.

rest of her career. When her daughter, Mary Catherine Bateson, was born in 1939 (her father was Mead's third husband, Gregory Bateson), Mead breastfed on demand, as she had seen women of the South Seas do, and found a paediatrician who supported her – the enormously influential Dr Benjamin Spock, for whose work she was a collaborator and champion. Between them they popularised natural childbirth and breastfeeding, both rarities in twentieth-century Western culture.

On their next extended field trip in the early 1930s, Mead and Fortune returned to New Guinea; this trip would prove to be as personally turbulent as it was creatively stimulating. Another anthropologist, Gregory Bateson, was working nearby. Mead told Benedict he had an air of 'vulnerable beauty' despite his lumbering six-foot-five frame and piercing intellect. When he and Mead fell in love, it sparked months of drama played out on sunny lagoons and in rain-soaked tents. Night after night they debated what to do in anguished gin-fuelled discussions by the dim light of kerosene lanterns and, at the end of it, having concluded that monogamy was not for her and decided to leave Fortune for Bateson, Mead realised something fundamental that would form the theme of her next book, *Sex and Temperament in Three Primitive Societies* (1935).

She had been examining three tribes in New Guinea with divergent attitudes to gender. Mead classified the peace-loving Arapash as maternal and feminine in their approach to life, although both sexes were involved in child-rearing and divided

Mead playing with local children in Manus Island during her research for
Growing Up in New Guinea *(1930).*

labour equitably. The Mundugumor, a head-hunting tribe, brought their children up with gender roles sharply divided and had a complex and brutal system whereby the men of the family could trade in their wives or sisters for another woman. Finally, the Tchambulis, or lake-dwellers, were led primarily by women, who provided for their families as well as nurtured them; their men lived in 'ceremonial houses' and spent their days in recreational activities and adornment. They put on 'grand masked balls where men dressed as women and women pantomimed intercourse'.

Mead compared how the three societies 'have dramatised sex-difference', illuminating 'what elements are social constructs', irrelevant to biological gender discrimination. Where the behaviour, and even the temperament, of men and women was seen to be sharply differentiated – as in 'our own society' – Mead looked at new research into 'masculine' women and 'feminine' men like the *berdache*, or inverts, of the Dakota Indians, similar to the Zuñi *ta'mana*, or man-woman. These Plains tribes exuded tough virility but their *berdache* wore women's clothes and did women's work without being valued any less. Citing Benedict's *Patterns of Culture*, Mead noted that recent studies had made Americans 'more sophisticated. We know that human cultures do not fall into one side or the other on a single scale and that it is possible for one society to ignore completely an issue [such as the treatment of old people and children, of men and women, of work and ritual] which two other societies have solved in contrasting ways.'

She observed that children develop not just as members of their tribe but as members of their gender. If they don't have the characteristics they are supposed to have – arrogance and aggression in the Mundugumor boys, for example, or even-temperedness among the Arapash – they become anxious and confused and both the individuals and their societies are diminished. She argued that each child, wherever he is born, should be encouraged to grow 'on the basis of his actual temperament' rather than according to gender-defined patterns of behaviour.

A richer, more expressive society would result. Instead 'of being forced into an ill-fitting mould … we must recognise the whole gamut of human potentialities, and so weave a less arbitrary social fabric, one in which each diverse human gift will find a fitter place'.

Like Benedict, her great love, and also probably like Edvard Westermarck and possibly William Rivers, Margaret Mead was deeply personally involved in this argument. It mattered to her and not just on a philosophical, theoretical or scientific level. She knew herself to be neither monogamous nor heterosexual at a time when women were expected to be both and she wanted to live in a world where differences like the ones she felt to be true were accepted, even valued, rather than ignored or, still worse, persecuted. Her work showed her that sexual identity was an artificial distinction. Echoing Boas's arguments about race, she sought to demonstrate that temperamental variations between individuals of the same gender were greater than variations between genders.

Boas and Malinowski, both normatively heterosexual, had paid scant attention to homosexuality in their research. In Samoa, Mead noted casual homosexuality between boys but less between men. A young man of twenty who had made advances towards other adult men was considered 'an amusing freak' by the girls but seen with 'mingled annoyance and contempt' by those who had rejected his overtures. She did not mention homosexuality between women, presumably because she did not observe it. In the main she found Samoa a tolerant and pragmatic society in which very little (at least relating to sex) was entirely taboo. 'Onanism, homosexuality, statistically unusual forms of heterosexual activity, are neither banned nor institutionalised. The wider range which these practises give prevents the development of obsessions of guilt' so prevalent in the United States. The 'acceptance of a wider range as "normal" provides a cultural atmosphere in which frigidity and psychic impotence do not occur'.

In 1935, Mead had encouraged Ruth Benedict to write an account of her life, slightly disingenuously explaining later that 'life stories were becoming a matter of anthropological interest'. Mead and Benedict's critics would have found this unforgivably 'psychological' but for Mead, life and work were intimately intermingled. Her three husbands, as well as the two women with whom she had serious relationships – Benedict until her death and, thereafter, Rhoda Métraux, with whom Mead lived from 1955, after Gregory Bateson left her, until her own death in 1978 – were all anthropologists, as was the daughter she had with Bateson; much of her work was fundamentally collaborative. The example of her life, as well as her outspoken stance on sex and gender, made her work political. Like Benedict, she was a supporter of 'free love' – a loose movement that supported freedom from state interference in personal relationships and economic as well as sexual liberation for women, rather than promiscuity. Women should be able to leave their husbands if they were unhappy, live with the person they loved without needing official sanction, be able to earn their own money and rely upon themselves; sexual pleasure should be mutual, not one-sided.

Where things became more complicated for Mead, foreshadowing some of the complexities of the developing feminist movement, was her belief that mothers were best fitted to provide the loving, nurturing environments children need. As Betty Friedan argued in *The Feminine Mystique* in 1964, specifically citing Mead, functionalist anthropology could be harnessed to demonstrate that a woman's indisputable biological role was as wife and mother. Friedan argued that Mead, with her blazing independence, might have been an inspiration for the feminist movement but all too often slipped back into glorifying and idealising motherhood as the central feminine role. Although Mead may have hoped that modern women would choose natural, loving motherhood, Friedan saw her attributing 'sexual specialness to everything a woman does'.

The dart may have stung but Mead's career was the best refutation of Friedan's critique. Over her working life she published more than 1,500 books and articles, many in the popular press. For sixteen years she had a column in the million-selling women's magazine *Redbook*; she was lauded by *Time* magazine and consulted by the US government. She was interested in everything that touched on modern life, from the space race to the early environmental movement to drug use to nuclear policy. Her great skill, apart from her optimism and approachability, was her quotability: 'Always remember that you are absolutely unique. Just like everyone else'; 'Laughter is man's most distinctive emotional expression'; 'I do not believe in using women in combat, because females are too fierce'; 'Every time you liberate a woman, you liberate a man'; 'Children must be taught how to think, not what to think.'

The Christian surgeon Paul Brand, in his 1980 memoir *Fearfully and Wonderfully Made*, remembered Mead, at a lecture, discussing the earliest sign of civilisation. It was, she declared, not tools, not agriculture, not pottery or iron, but a healed femur bone, which she held up before the audience: 'The healed femur showed that someone must have cared for the injured person – hunted on his behalf, brought him food, and served him at personal sacrifice. Savage societies could not afford such pity.' Compassion, perhaps love, was what made us human, was the message.

In 1939, having been in the public eye for over a decade, Mead observed that the questions being asked by her audience were different from the ones she heard when her first book came out. The man in the street, she said, had become conscious of 'the creaking and groaning of the social structure' and the thing he wanted to know was 'When we build a new world, what kind of world do we want to build?' Mead was certain she knew the answer: 'Primitive man, secure in a closed and ordered universe, has a dignity that we have lost.' He has few doubts and confusions, but almost no opportunity to change the fate his society has mapped out for him. But is his homogeneity – or totalitarianism, in Mead's word – bought at too high a price? Even if we wanted to return to that simple dignity, it would be impossible. Throughout her work, Mead argued instead for a new world in which the 'myriad gifts' of humanity can intermingle, 'a world of interrelated and integrated values which will replace both the homogeneity of the savage and the confused and frustrated heterogeneity of the twentieth century'.

CHAPTER NINE

Insider/Outsider

Zora Neale Hurston in New Orleans, 1928

—— • ——

When Zora Neale Hurston walked into one of Franz Boas's Barnard lectures in the autumn of 1925, he saw immediately that she was the recruit he needed for one of his dearest-held anthropological objectives. His career had been marked by measuring skulls – even stealing them from cemeteries, in his worst moments – dating back to his earliest work in Berlin in the 1880s with the physical anthropologist Rudolf Virchow, an expert in scientific craniometry; when he was at Clark University in the 1890s, he had been pilloried by the local paper for measuring the skulls of schoolchildren. Now he needed help measuring the heads of African Americans — then in a migration pattern from very poor rural Southern areas to more prosperous urban Northern ones — to contribute to his study of whether people's skulls changed shape and size when their environment changed, but it was awkward for white academics to stop black people in the street and request this. One of the things William Rivers had cautioned students of ethnography about was how offended 'native' people could be by being measured but, wherever they came from, very few people wanted to be examined as the subjects of experiments.

This was where Hurston came in. With her irrepressible smile, she was impossible to refuse or intimidate. 'When I set my hat at a certain angle and saunter down 7th Avenue [Harlem's main drag] ... the cosmic Zora emerges,' she wrote. Persuading people to let her measure them would be easy. 'How *can* any deny themselves the pleasure of my company? It's beyond me.' She had arrived in New York that January with $1.50 in her pocket and a blazing desire 'to get *All*' from life. At thirty-four years old – though she claimed to be twenty-four – she had come a long way from Notasulga, Alabama.

Hurston's childhood had been mostly spent in Eatonville, Florida, America's first incorporated black town, where her father had been pastor in one of the Baptist churches and mayor. Hurston was thirteen when her mother died. Her father and stepmother sent her to a segregated boarding school in Jacksonville, where she discovered she was black: her first day there was 'the very day that I became colored'. In 1917, aged twenty-six, she reinvented herself as a sixteen-year-old to qualify for the free high school education offered by the all-black Morgan College in Baltimore, Maryland, before moving to Washington, DC, where she attended another all-black college, Howard University, co-founding the student newspaper there.

She dreamed of being a writer and the short story she wrote in 1921, during her time at Howard, 'John Redding Goes to Sea', brought her membership of Alain Locke's literary club, the Stylus. It was probably this introduction, through Locke, the first African American Rhodes scholar and a writer and philosopher, to what would become known as the Harlem Renaissance, that inspired Hurston to move to New York. 'Drenched in Light', the first story she published nationally, came out in *Opportunity* magazine in December 1924; the central, autobiographical character was called Isis Watts.

Still with ten years sliced off her age, Hurston hustled – a word she used a lot in her letters – to get a scholarship to Barnard, supported by the novelist, suffragist and Barnard trustee, Annie Nathan Meyer. She was the sole black student there; later she would say, 'I feel most colored when I am thrown against a sharp white background.' 'I suppose you want to know how this little piece of darkish meat feels at Barnard,' she wrote to a friend after her first semester. 'I am received quite well. In fact I am received so well that if someone would come along and try to turn me white I'd be quite peevish at them.'

Despite her warm welcome, there were disappointments and humiliations, too. At first 'the girls' had urged her to come to the junior prom, promising that if she brought a date 'as light as myself' they would exchange dances with her. It's hardly a wonder they wanted her, since 'when Zora was there, she *was* the party', but Meyer, her patron, thought attending would be inappropriate. Still her spirit was undaunted, as she told Meyer: 'I constantly live in fancy in seven league boots, taking mighty strides across the world, but conscious all the time of being a mouse on a treadmill ... The eagerness, the burning within, I wonder the actual sparks do not fly so that they may be seen by all men.'

At Barnard, she enlisted on a course in anthropology. For a person who later said she was born 'a child that questions the gods of the pigeon-holes', it was an obvious fit. By the spring of 1926, she was being trained up in anthropometry by Melville Herskovits, one of Boas's colleagues, 'as Boas is eager to have me start'. This initial anthropological work – her first fieldwork, in a way – was standing on a street corner in Harlem with a pair of calipers, stopping passing families and asking to measure their heads. As Boas would later put it in the inevitable foreword to her first book, her charm and ability, as an African American herself, to win the trust of 'the companions of her childhood' made her the perfect collector of African American material. It was 'a Job for the Summer', with the promise of more to come 'if I make good'. Elsie Clews Parsons, another enthusiast of 'Negro' folklore, was also dangling offers in front of her and Hurston was determined to be quick and accurate for the 'glorious career before me'.

The following year Boas helped Hurston get funding ($700 from the Association for the Study of Negro Life and History and $700 from Parsons at the American Folklore Society) to return to her home state of Florida to gather 'materials dealing with the traditional beliefs, legends, sayings and customs of blacks' and, implicitly, to demonstrate their richness and beauty. Hurston's background, almost unique among her peers, would help her fully understand and convey to white America 'the character of American Negro life'; having received the civilising benefits of education, she was being sent back into her own world to record and analyse it.

The study of African American culture was seen partly as salvage anthropology, trying to reconstruct how past or vanishing cultures had functioned – in some areas it was thought to be as 'primitive' as that of any South Seas tribe – but Hurston was among the first of her colleagues also to investigate contemporary culture through 'the spy-glass of anthropology'. 'Discovery 7', she wrote as a postscript in a letter of April 1928 from Florida, after six other 'discoveries': 'Negro folk lore is *still* in the making [...] a new kind is crowding out the old.'

She set off in March 1927, proudly at the wheel of a newly bought second-hand car, a Nash coupé she named Sassy Susie, and almost at once had to write an explanatory letter to Franz Boas, who had been contacted by the credit company checking up on the references given by this anomalous single black woman somehow able to afford monthly payments on a car. He was worried she was borrowing money; she had to explain that she needed the car for her expedition, travelling round the South with a typewriter and great piles of notes and carbon paper. 'I decided to do as Dr [Gladys] Reichard had done – and buy an old one cheaply,' she

Zora Neale Hurston in Florida, c. 1928, with her coupé, nicknamed Sassy Susie. If she couldn't find a hotel room she slept in the car, with her pistol handy.

'I love myself when I am laughing. And then again when I am looking mean and impressive,' Hurston told her friend Carl Van Vechten.

told him, adding tantalisingly that she had arrived just in time, for 'the negro is not living his lore to the extent of the Indian. He is not on a reservation, being kept pure. His negroness is being rubbed off by close contact with white culture.'

'Folklore is not as easy to collect as it sounds,' Hurston observed in the introduction to *Mules and Men* (1935), her ethnographic account of her home town, Eatonville, Florida. 'The best source is where there are the least outside influences and these people, being usually under-privileged, are the shyest. They are most reluctant at times to reveal that which the soul lives by. You see we are a polite people and we do not say to our questioner, "Get out of here!" We smile and tell him or her something that satisfies the white person because, knowing so little about us, he doesn't know what he is missing. The Indian resists curiosity by a stony silence. The Negro offers a feather-bed resistance. That is, we let the probe enter, but it never comes out. It gets smothered under a lot of laughter and pleasantries.' This was perhaps the first time that anthropologists had heard, from one of their own number, what it felt like to be the object of anthropological research.

On and off through 1927, 1928 and 1929, Hurston drove across the Southern states, mostly alone except when she could persuade a friend to join her, staying in 'colored' hotels or, when a room couldn't be found or she was routed by bedbugs, sleeping in her car with a little chrome-plated pistol on the seat beside her. She bumped along dirt roads criss-crossed with palmetto roots, 'but the little old car gets me anywhere'. Her research took her from Eatonville to the fringes of black life, the railroad camps, turpentine stills, sawmills and phosphate mines where the poorest labourers worked. During the day she took down their work songs, spirituals and stories – or 'lies', as they called them – paying for the best ones; at night, she drank, sang and danced with them in jooks, roadside dance halls. 'Wish you were here,' she wrote mischievously to the poet Langston Hughes from Miami, 'because such good likker can be had.' (This was Prohibition, so the liquor would have been illicitly imported Caribbean rum.) She collected lyrics for Porter Grainger, a Harlem pianist and songwriter, and in 1928 received royalties for a song called 'Jelly Roll' – perhaps the version sung by Bessie Smith, with whom Grainger worked in the 1920s and a friend of Hurston's, too.

Although she was, as she told one of her Harlem friends, 'lost to Bohemia forever. No more parties – just work and *work*', Hurston revelled in her road trips. 'Florida: gorgeous sunlight, fleas, flowers, frogs, ferns, alligators, poincianas, flies, cypress, roses, magnolia, roaches, bougainvillea, gnats, pines, roaches, china berry trees, fleas, bedbugs and magnolias amid dazzling palms and stretches of water,' she wrote to another friend. 'Wish you could see it all.' She kept in touch with literary friends Alain Locke, Countee Cullen and Carl Van Vechten, as well as the anthropologists Franz Boas, Melville Herskovits and Ruth Benedict.

Somewhere along the way, in May 1927, she married Herbert Sheen, a jazz musician she had known since her Howard University days five years earlier, who was studying to be a doctor in Chicago.

But it would have been hard for a husband not to be a hindrance. Within a year, she had decided to let him go, telling Hughes, 'I am going to divorce Herbert as soon as this is over. He tries to hold me back and be generally obstructive.' They were officially divorced in July 1931: 'I hear that my husband has divorced me, so that's that. Don't think I am upset, for your lil Zora is playing on her harp like David. He was one of the obstacles that worried me.'

These two letters about her marriage were to Hurston's two most important correspondents in this period, respectively Langston Hughes and the woman she called 'Godmother', Charlotte Osgood Mason. Hughes was her 'mainstay in all crises. No matter what may happen, I feel you can fix it. Let me hear you soon, honey,' she ended one letter. 'Lovingly yours, Zora. Do you need some money?' Like her, Hughes was a member of the 'Niggerati' – a phrase coined by Hurston to describe the free-living young artists of the Harlem Renaissance, each animated by a spirit 'of outrageous, amoral independence'. His heart beat to the sound of the tom-tom, according to the painter Richard Bruce Nugent, who wore a single gold earring and shared 'Niggerati Manor' with the poet Wallace Thurman and Thurman's white lover. Nugent had decorated the walls with bright-coloured penises; he joked that gin flowed from the taps. 'Nobody was in the closet,' he said later. 'There wasn't any closet.' He thought Hurston was 'one of the most alive people' he knew.

With Charlotte Osgood Mason, Hurston played the grateful 'pickaninny' rather than the untamed spirit. Hughes, another of Mason's black protégés, described the relationship with the bittersweet benefit of hindsight: 'In her youth she [Hurston] was always getting scholarships and things from wealthy white people, some of whom simply paid her just to sit around and represent the Negro race for them, she did it in such a racy fashion. She was full of side-splitting anecdotes, humorous tales, and tragi-comic stories, remembered out of her life in the South as a daughter of a travelling minister of God. She could make you laugh one minute and cry the next. To many of her white friends, no doubt, she was a perfect "Darkie", in the meaning they give the term – that is a naïve, childlike, sweet, humorous, and highly coloured Negro.'

Mason was a rich widow and patron of the arts with a passion for the 'primitive' and a corresponding desire to dominate the young black talents she supported, who literally sat on stools at her feet when they visited her Park Avenue apartment. She helped and encouraged Alain Locke, Miguel Covarrubias (artist, illustrator and ethnologist) and Arthur Fauset (the only other black ethnographer in Harlem, whose novelist sister, Jessie, was literary editor of the NAACP's magazine, *The Crisis*). Mason's three-year backing of Langston Hughes ended in 1930, when he rebelled against her dictum that he correspond only with her and read and listen only to what she suggested.

In September 1927, towards the beginning of his association with Mason, Hughes introduced her to Hurston, and on 1 December the two women signed a contract that provided Hurston with a car, a camera and $200 a month, for which she was to collect and write up material concerning 'music, folk-lore, poetry, hoodoo, conjure, manifestations of art, and kindred matters existing among the American negroes' that Mason 'because of the pressure of other matters' was unable to collect herself. In return, Hurston promised not to use any of this material in any way; legally it was Mason's. The contract would run for five years.

Hurston understood what was required of her and always insisted that her devotion to Godmother was genuine, despite the pose of adorable supplication it required as well as the sacrifice of her own literary and academic ambitions. Her first surviving letter to Mason thanks her for 'the happiest Christmas [1927] of all my life ... Godmother dearest, far-seeing one, you have given me my first Christmas ... when reality met my dreams. The kind of Christmas that my half-starved childhood painted.' She itemised every penny she spent of Mason's, down to sanitary towels, and agonised over unintentionally offending her. Mason was not the only white patron on whose support, tangible and intangible, Hurston had relied – without Meyer at Barnard and Meyer's friend, the novelist Fannie Hurst, for whom Hurston worked briefly as a disorganised but amusing secretary, she would not have made it through her first year in New York – but she was the most important of her 'Negrotarians', as she called them behind their backs.

Carl Van Vechten was another rich white New Yorker, a critic, journalist, photographer and novelist who collected intelligent, stylish black artists like Hurston, supporting them not in cash like Mason but in friendship, admiration and promotion. When he sent Hurston some pictures he'd taken of her, she replied, 'I love myself when I am laughing. And then again when I am looking mean and impressive.' He adored African American culture, believing its spirit, humour and energy was the very essence of modern America, and he sang the praises of Hurston, Hughes, Bessie Smith and their friends, joined them in Harlem speakeasies, threw parties for them and wrote up their work for the white readers of *Vanity Fair* and *The New York Times*.

Another group of interested patrons, though perhaps they would not have thought of their relationship with Hurston like that, were the anthropologists of Columbia University. Hurston received her BA in anthropology from Barnard in 1928, when she was thirty-seven, and spent the next two years doing fieldwork as a graduate student at Columbia (slightly clandestinely, as she was at the same time the 'agent' of Charlotte Mason). Occupying a unique position on one hand as an actual member of a society considered exotic and primitive enough to merit ethnological investigation and on the other trained to be one of those investigators, she personified the 'double-consciousness' W. E. B. du Bois had defined in 1897 as 'always looking at one's self through the eyes of others, or measuring one's soul by the tape of a world that looks on in amused contempt and pity'. But, as Hurston herself would observe, 'all clumps of people turn out to be individuals on close inspection'.

Still, there was something in the adjectives her colleagues used about her that must have hurt or enraged her, or both, inextricably mixed with her gratitude at working with them and her desire not to upset the career she'd struggled to achieve. Boas praised her in the foreword to her first book, the ethnographic study *Mules and Men*, for her lovability – not exactly the scholarly description he would have used for Robert Lowie, Alfred Kroeber or even Ruth Benedict. Elsewhere, off the record, he commented that he found her a 'little too much impressed with her own accomplishments'. After her first stint in Florida in 1927, Boas sat her down and told her to be more systematic in her approach to note-taking, ensuring that the information she gathered could be placed into an academic context; she left the meeting in tears. It didn't help when she started working for Mason, who wanted authentic, unadulterated material, a contrast to the rigour and restraint demanded by anthropology.

But Boas supported Hurston when funding for a PhD was withdrawn in the early 1930s, apparently because she was seen as ill-disciplined and unreliable, and he and Benedict acted as referees when she successfully applied for a Guggenheim Fellowship in 1936 (Margaret Mead had been turned down the previous year). Among other things, this allowed her to research what would become her great novel, *Their Eyes Were Watching God*, in Jamaica and Haiti, a reflection of what she once said she wanted to achieve in her work, 'literary science'.

The line between the academic world she never quite felt a part of and the 'Negro' world from which she had come was a treacherous one to navigate. Her work wasn't easy, despite her connection with it. 'You see there are Negroes with "Race

Consciousness" and "Race Pride" drilled into them and they resent any thing [*sic*] that looks like harking back to slavery,' she told Benedict, relaying what some of her potential subjects might say to her and each other when she began interviewing them. "'Whuts all this fur, nohow? Some white folks is trying tuh get us something so they kin poke fun at us ... They wants tuh drag us down an y'all sittin here like a pack uh dunces pewking yo' guts.'"

She seems to have found it hard not to adopt a subordinate, almost childlike persona with people she saw as white 'patrons'. When she persisted in calling the head of the Guggenheim Foundation, Henry Allen Moe, 'bossman' and asking for his advice, he replied that he was at her service: 'I am not *busha* [Jamaican for boss]. The Fellows are *busha*. The Foundation exists for them. So, *busha* (or whatever the feminine is), don't hesitate please, to let me know what's what and what you want.'

In December 1927, Hurston began interviewing Cudjo Lewis, an eighty-six-year-old man who was the last known survivor of the last American slave ship. Born in 1841 among the Isha Yoruba people of West Africa, Lewis – whose African name was Oluale Kossola – went through the training of a young soldier and was about to undergo the initiation preparatory to marriage in 1860 when the female warriors of Dahomey stormed his town, massacred his people and marched the survivors to the barracoons (slave enclosures) in Ouidah. There they were sold to a white slave trader from Alabama, who in the end left without paying for his (by then illegal) human cargo. Alongside more than a hundred other captives, young men and women, Lewis was taken naked aboard the *Clothilda* and sailed across the Atlantic. In Mobile, Alabama, he was bought by two brothers for heavy work, loading and unloading cargo going up and down the river to Montgomery, until he was freed by Yankee soldiers in 1865.

Hurston spent several months with Lewis, visiting him regularly with presents of peaches, Virginia hams and insect powder, building a genuine friendship. Sometimes he wanted to talk and they sat on his porch together eating watermelon; sometimes he ignored her or sent her away; sometimes he asked her to take him into town. Once they got blue crabs from the bay and the next day they ate them together. Through him she could trace the journey he and so many others had made 'from humanity to cattle'; this was personal to her, because all four of her grandparents had been slaves. When she photographed Lewis, he wore his best suit, barefoot: 'I want to look lak I in Affica, 'cause dat where I want to be.' He asked her to photograph him in the cemetery, by the graves of his wife and children.

Hurston published an article about Lewis in 1928 and, in 1931, her manuscript *Barracoon: The Story of the Last 'Black Cargo'* was ready, dedicated 'To Charlotte Mason, My Godmother, and the one Mother of all the primitives, who with the Gods in Space is concerned about the hearts of the untaught.' Mason's money had made Hurston's work with Lewis possible and she had been much in Hurston's thoughts during her time with him; she sent her a melon and told Lewis all about 'the nice white lady in New York who was interested in him'. (Through Hurston and afterwards directly, Mason also sent money to Lewis.) But Viking Press, which had expressed an interest in publishing the book, requested that Hurston change the narrative from Lewis's vernacular dialect to 'language' and Hurston, encouraged by Mason (or obedient to her wishes), refused. The manuscript would lie neglected until 2018, when it was published with a foreword by Alice Walker, who hailed it a 'maestrapiece'.

When Hurston wrote to Franz Boas during this period, she had to be careful not to cross the line she had agreed with Mason about the material she was gathering. Replying to his first letter in December 1928, she told him how proud she was that he'd written and how much she hoped her work on sympathetic magic and Creole languages would please him but that a promise had been 'extracted' from her not to write to anyone. 'Of course I have intended from the very beginning to show you what I have, but after I had returned. Thus I could keep my word and at the same time have your guidance. I am finding a lot of things which will intrigue you.'

Her letters reveal how hard she was working. In August 1928, having finished her interviews with Cudjo Lewis, she moved to New Orleans. 'Dear Langston,' she wrote, 'I have landed here in the kingdom of Marie Laveau˙ and expect to wear her crown

---

* Marie Laveau was the legendary Voodoo Queen of New Orleans. Descended from African, French and Native American people, she was born in 1801 and died in 1881 after a long career as a herbalist, midwife and voodoo practitioner.

someday—Conjure Queen as you suggested.' She found herself a three-room house in Algiers, across the river from New Orleans as Brooklyn is from Manhattan, which cost $10 a month. It had electric light and running water and was situated in what she described as 'a splendid neighbourhood from the point of view of collecting materieal [*sic*]' – historically poor, black and proudly independent. Slaves had been unloaded and held here before being taken over to the city for sale in the eighteenth century; in the nineteenth century, an eccentric slave-owning benefactor had donated land for freed slaves to live in. Jazz and hoodoo, a Creole voodoo particular to New Orleans as opposed to the Caribbean, flourished here. In November, she was still 'knee deep in it'; by the following April she was 'just beginning to hit my stride'; six months later she talked of having worked 'two years without rest'. That was October 1929, the month of the Great Crash.

Hurston's fieldwork was more participant than observer. She immersed herself entirely into everything, anthropological or not, and hoodoo was no exception. In her memoir, *Dust Tracks on a Road* (1942), she described meeting the Devil at midnight at a crossroads, and her initiation into the practices of various 'doctors' with whom she studied. For one, 'I lay naked for three days and nights on a couch, with my navel to a rattlesnake skin which had been dressed and dedicated to the ceremony. I ate no food in all that time. Only a pitcher of water was on a little table at the head of the couch so that my soul would not wander off in search of water and be attacked by evil influences and not return to me. On the second day I began to dream strange exalted dreams. On the third night I had dreams that seemed real for weeks. In one, I strode across the heavens with lightning flashing from under my feet, and grumbling thunder following in my wake.

'In this particular ceremony, my finger was cut and I became blood brother to the rattlesnake. We were to aid each other forever. I was to walk with the storm and hold my power, and get my answers to life and things in storms. The symbol of lightning was painted on my back. This was to be mine for ever.'

Most terrifying of all 'was going to a lonely glade in the swamp to get the black cat bone. The magic circle was made and all of the participants were inside. I was told that anything outside that circle was in deadly peril. The fire was built inside, the pot prepared and the black cat was thrown in with the proper ceremony and boiled until his bones fell apart. Strange and terrible monsters seemed to thunder up to that ring while this was going on,' she wrote, not quite able to achieve the scientific detachment her discipline required. 'It took months for me to doubt it afterwards.' Later, asking Boas for advice on writing up her material, she had regained her educated scepticism. 'Is it safe for me to say that baptism is an extension of water worship as a part of pantheism just as the sacrament is an extension of cannibalism? ... May I say that all primitive music originated about the drum?'

Hurston found that belief in and practice of hoodoo had been at its peak in the 1880s, so much of her research in New Orleans was salvage anthropology. As a student of contemporary African

American life and culture, as well as a natural storyteller, she risked being labelled a populariser, repackaging what was called Negro folklore for new audiences in theatres, at speakeasies and in galleries; that was why her academic work was so important. African American music, dance, sculpture, textiles and stories, uncovered by anthropology, were seized upon by a new generation eager for ways to shape and take pride in their ethnic identity. When anthropology demonstrated the unique value of black culture, for example in the 'Negro'-themed issues of the journal of the American Folklore Society, it made demands for racial equality all the more powerful; as an anthropologist with degrees from Barnard and Columbia as well as a Guggenheim Fellowship, Hurston and her subject matter demanded to be taken seriously.

Some white Americans found some elements of black culture distasteful, even frightening – crude, rude, illogical, opaque – but Hurston wanted to show it all as part of a whole. She frequented the jooks on the shabby outskirts of town as well as the glitzy parties given at Harlem's Dark Tower by the hair-straightening heiress A'Lelia Walker, daughter of the first African American millionairess, and the literary salons of Alain Locke; she loved the dirty innuendo of the blues as much as the lofty power of spirituals. Just as African Americans needed to know the hard truth revealed by Cudjo Lewis's memories that Africans had been sent to America partly by other Africans, so too was it important to seek out the authentic in this new world, not to pasteurise it for different audiences, not to choose some parts and ignore others.

Only then would it be seen and valued for itself. 'It makes me sick to see how these cheap white folks are grabbing our stuff and ruining it,' she wrote to Hughes in the late 1920s. 'I am almost sick – my one consolation being they never do it right and so there is still a chance for us.'

Hurston carried on working hard through the Depression, but academic funding (especially for untenured academics, who of course tended to be women) was much harder to come by when money was so tight. Even so, she celebrated being released from her association with Mason in 1933, telling Benedict after their contract ended that December, 'I have kicked loose from the Park Avenue dragon and still I am alive! I have found my way again.'

She continued her research, especially in the Caribbean; she taught at various colleges, establishing a school of drama 'based on pure Negro expression' at Bethune-Cookman College in Daytona Beach, Florida; she hunted down folk songs; she wrote plays, falling out spectacularly with Langston Hughes over a collaboration called *Mule Bone*, which would not be performed in either of their lifetimes. Memorably, she described the thrill she felt when she heard from Lippincott that they wanted to publish *Mules and Men* (1935), the literary-ethnological collection on which she had worked for five years, as well as a first novel, *Jonah's Gourd Vine*: 'You know the feeling when you found your first pubic hair. Greater than that.' On the whole, though, throughout her career, 'book sales, grants and fellowships [were] too few and too paltry, ignorant editors and a smothering patron [Mason]' restricted her output.

To paraphrase Langston Hughes, if the 'Negro' had been in vogue during the 1920s, from 1929 onwards that moment of cultural openness was over, and with it Hurston's time in the sun. The final decades of her life were marked by increasing struggle, disappointment and despair: in poor health, intermittently homeless and strapped for money (at one stage she had to pawn her typewriter), the object of spiteful false accusations, unhappy in love and friendship and, most cruelly, her work underappreciated. But she never saw herself as a victim, vehemently opposing preferential treatment for African Americans: 'If I say a whole system must be upset for me to win, I am saying that I cannot sit in the game, and that safer rules must be made to give me a chance. I repudiate that. If others are in there, deal me a hand and let me see what I can make of it, even though I know some in there are dealing from the bottom and cheating like hell in other ways.' This emphasis on the individual and her feeling that left-wing politics like the New Deal whipped up 'racial antagonism' set her apart from her left-leaning friends and academic colleagues. Defiantly independent, proudly anti-communist, she insisted, 'I belong to no race nor time.'

Although much of her work, especially her early work, centred on the discovery and appreciation of African American culture, she didn't want to be defined by her African Americanness: she wanted to write novels about white people as well as black people, to study whatever cultures and personalities sparked her interest rather than just those about which other people assumed she would have insights. To the chagrin of many of her friends, she refused to be an activist for civil rights as the movement reached a climax during the 1950s. The leading characters of her last published novel, *Seraph on the Suwanee* (1947), were white, a daring liberty for a black author at this time; during the final decade of her life, the project obsessing her was a life of Herod, the link between Judaism and Christianity, whom she believed to have been a great leader.

Work remained her solace and inspiration, 'the nearest thing to happiness that I can find'; generosity of spirit the key to a new world in which race would be irrelevant. 'I have no race prejudice of any kind,' she wrote at the conclusion of her 1942 memoir, *Dust Tracks on a Road*. 'My kinfolks, and my "skinfolks", are dearly loved ... But I see their same virtues and vices everywhere I look. So I give you all my right hand of fellowship and love, and hope for the same from you. In my eyesight, you lose nothing by not looking just like me ... Let us all be kissing-friends. Consider that with tolerance and patience, we godly demons may breed a noble world in a few hundred generations or so.'

CHAPTER TEN

The Bluestocking

Audrey Richards in Zambia, 1930–1931

——•——

In May 1930, aged thirty-one, Audrey Richards embarked on the long journey from London to the highlands of what was then north-eastern Rhodesia in modern Zambia. Under the guidance of her LSE supervisor, Bronislaw Malinowski, she had planned a year's fieldwork with the Bemba (or Babemba) people. Times had changed since an anthropologist might set off to the edge of the known world laden down with crates of exotic foods and custom-made tropical outfits; in 1930, Richards travelled rather more lightly than Malinowski had in 1913, taking just a tent and a bicycle alongside her Brownie camera and the latest Bemba dictionary.

She had not remembered everything: as she boarded her train at Victoria, the first leg of a weeks-long voyage, Malinowski ran down the platform towards her with some last-minute advice, bringing ('to my great regret', she remembered with a smile, many decades afterwards) coloured pencils for notes ('brown for economics, red for politics, blue for ritual') and coloured cards for linguistics ('yellow for adjectives, blue for so-and-so, red for proverbs'). 'We were all trying to be very scientific,' she said later. 'That was the great craze of the time.'

Africa, Richards wrote in one of the articles she published on her return, presented the ethnographic fieldworker with new challenges. While the established British anthropologists of the generation above her, including Malinowski and William Rivers, had studied smaller, more static societies, often isolated from contact with the outside world, African tribes were scattered over large areas with rapidly increasing, rather than decreasing, populations. Most crucially, contact with Europeans (dating, in the case of the Bemba, from less than fifty years earlier) was transforming African society at every level. Richards found Africa 'at a moment of dramatic and unprecedented change in tribal history' that was, 'in effect, a revolution'. It was hard to keep up: 'the whole picture of African society has altered more rapidly than the anthropologist's technique'.

In the tradition of well-connected British travellers, she found friends nearby with whom to base herself, at the newly built Italianate villa of the monocled grandee Sir Stewart Gore-Browne. Shiwa Ngandu, or Lake of the Royal Crocodiles, was

---

* Gore-Browne and Shiwa Ngandu are the subjects of Christina Lamb's 1999 book *The Africa House*. Introducing *Black Heart: Gore-Browne and the Politics of Multi-Racial Zambia* (1977) by Robert Rotberg, Kenneth Kaunda (the first president of Zambia) praised him as 'one of the rare white people at the height of Zambia's battle for national independence who stood on the black man's side to struggle shoulder to shoulder for

an estate of 22,000 acres, the focal point of social life in the remote Zambian plateau, peopled on and off with British and African guests, black and white, enjoying the generous hospitality of their eccentric but benevolent host and his spirited young wife.

But Richards didn't linger to luxuriate in cocktails and European plumbing. She was, as one colleague observed, 'a born fieldworker', who could easily be recognised in the villages she visited 'by the long train of girls and boys at her heels, the foremost proudly carrying her notebook and pencil' – no doubt those coloured crayons and cards. She also developed proper friendships with the people she worked with. Paul Mushindo, a local Presbyterian minister who assisted her, said she was like a sister to him. She 'thought I was helping her in her duties', he said, but 'I felt I was in a university for study'.

Although according to the Malinowskian convention she was accorded the status of chieftainess and used the royal Bemba dialect – which she found especially useful for taking population censuses – she also became fluent in the vernacular of the farm and village, with its rich taxonomy of mopane worms (edible caterpillars), boiled, dried and roasted. After a month's intensive study with the White Fathers, French Catholic missionaries who

freedom and justice … he heroically challenged the colonialist and racist structures of power and worked relentlessly for their destruction even at the risk of his own life'. Gore-Browne disliked Richards, whom he called 'that anthropologist woman with her clever talk and meddling ways', for helping his much younger wife Lorna realise she was happier away from Shiwa Ngandu, and him, than she was there.

had been in Zambia since the 1890s, it took four or five months in the field until she was fluent enough to gather information directly, especially from Bemba women, who tended not to speak any other language. Richards noted that no Rhodesian district commissioner in living Bemba memory could speak a Bemba dialect; before anthropologists arrived, only missionaries and settlers had taken the trouble to learn one.

She would spend three or more weeks in each village taking detailed statistical surveys, living in her tent – she liked to joke that 'it was easier to eavesdrop' through canvas. When she moved on to the next, she rode ahead on her bicycle (the vehicle of choice in these parts: even Gore-Browne went on safari by bicycle), followed by a 'motley procession' of bearers with 'a clatter of cooking equipment' tied together with bark and the kitchen boy bringing up the rear 'with a live chicken strung by its feet round the barrel of your rifle, and a couple of flat-irons in a basin on his head'.

Time was the anthropologist's great asset, combined with the lack of any agenda apart from the desire to observe. Missionaries and colonial officials might both become effective amateur ethnographers in their spare time but their *raison d'être* was altering the people they encountered, endeavouring with varying degrees of success to mould them (respectively) into Christians or taxpayers; the anthropologist was simply recording what she found, watching (as Richards put it) as well as listening, alert to the ripples of change. As a woman, no one ever imagined she was a colonial official; rather she was assumed to be a nurse and mothers would bring her their sick babies to cure.

Contact with Europeans was transforming the once-powerful Bemba tribes into an impoverished people. Unlike the native people of North America and Australia, who bitterly resisted what they were told was 'progress', the Bemba, Richards observed, were 'almost dangerously susceptible to the influence of Western civilisation'. But since they had no craft or profitable crop to trade and no way of storing their wealth, where for example the Kikuyu bred and owned cattle, the transition to a money-based, tax-paying economy was devastating. Their collective ambitions, once satisfied by tribal wars, had been replaced by the desire for European clothing, now their 'dominant craving' and their only means of gauging social status. Caught between two worlds, their villages and with them their social structures were crumbling: the white man, they told Richards, 'has taught us not to do work except for money'. The men of the tribe were disappearing to mine copper – more than half the adult male population were working in the mines of Zambia's Copperbelt district in 1930, sometimes walking a thousand miles to get there – leaving the tribe's villages and gardens desolate, inhabited only by women, old people and children.

Endlessly enthusiastic for her work, Richards found life in the field could be as exhausting as it was stimulating: 'There is the difficulty of taking photographs and simultaneously writing notes during rites that take place in bush and village and on the road between the two.' Ceremonies were often held late at night, in small huts of perhaps eight feet wide into which were crowded twenty or thirty people, well refreshed with beer, around a large fire. 'The observer is dead sober, nearly stifled, with eyes running from the smoke, and straining all the time to catch the words from the songs screeched around her, and to transcribe them by the firelight that penetrates occasionally through the mass of human limbs.' According to Lorna Gore-Browne, who accompanied her as translator and assistant on some of her expeditions, Richards never panicked. Pushing her auburn hair impatiently out of her eyes, she was always able to see the funny side of any small disaster. Her profound sense of duty was animated by an intense appreciation for the ridiculous and the incongruous and for the fallibilities of mankind – usually directed at herself.

Born in 1899, Richards came from a family enmeshed in the British establishment, not aristocrats but intelligentsia, whose members were as likely to shine in the Indian Civil Service as on the High Court circuit; one of her cousins, R. A. Butler, was both Chancellor of the Exchequer and master of Trinity College, Cambridge. Girls were expected to become capable matriarchs rather than academics or social workers, though, and it was despite rather than because of her parents that Richards read natural sciences at Newnham College, Cambridge. Inspired by Alfred Haddon and William Rivers, anthropology and social sciences were all the rage: even the classics department was shot through with 'Durkheimian fervour'* (a reference to the legendary French sociologist Émile Durkheim).

A spell teaching at her old school, Downe House, followed, as well as a useful period assisting the classicist Gilbert Murray in the analysis of ancient Greek ritual. In the early 1920s, Richards went to Frankfurt with the Friends' Ambulance Unit Family Welfare Settlement. This work, during which she went hungry as often as the families she looked after, steeped her in 'the experience of others' deprivation'; from this time onwards, nutrition would be a focus. Recommended to Bronislaw Malinowski at the LSE by the socialist political scientist Graham Wallas, the father of a Newnham friend, she began postgraduate work in 1927.

Both anthropology and the LSE were considered quite radical in the 1920s and 1930s but, as the social historian R. H. Tawney observed (and as Boas's school discovered at Columbia), 'it is agreeable to belong to an institution which does not inherit traditions but which makes them'. Anthropology was an intimate department and women were made welcome – Hortense Powdermaker was another student of Malinowski's in the 1920s, along with a Newnham contemporary of Richards, Lucy Mair; and Charles Seligman always worked closely with his wife, Brenda. Malinowski had none of the 'horror of the clever woman' so common at the time and 'women blossomed in this

---

* For example, the second book by Jane Harrison, fellow at Newnham and 'the cleverest woman in England', was *Thamis: A Study in the Social Origins of Greek Religion* (1912). See Mary Beard, *The Invention of Jane Harrison* (2000).

atmosphere of being taken completely seriously'. A sense of practical utilitarianism, that social sciences could be harnessed for social improvement, permeated the discipline.

Richards became an integral part of Malinowski's circle, close to his invalid wife, Elsie, who suffered terribly from multiple sclerosis, and a frequent babysitter for his daughters. Helena, the youngest Malinowski daughter, remembered that Richards was 'almost as great an influence on his life as he was on hers'. When Elsie died in 1935, they hoped Malinowski might marry her but, (reading between the lines) although she loved his girls, she seems to have been unwilling to give up a fascinating new career to become a housewife. In 1940, Malinowski married Valetta Swann, who was uninterested in his children, and Richards offered to become their guardian. Helena adored her, describing her years later cooking at the work-covered kitchen table of a tiny flat in Oxford using typing paper instead of rice paper, muttering, 'I must not make my meringues on my kinship tables!'

Following in Malinowski's footsteps, Richards' first book was largely based on the library research for her thesis. The introductory sentence of *Hunger and Work in a Savage Tribe* (1932) set the tone for her career: 'Nutrition as a biological process is more fundamental than sex.' At once tipping her cap to Malinowski and challenging his work, she argued that food was an individual's primary need and determined the nature and form of a group's activities and rituals; its very ubiquity rendered it unnoticeable, which perhaps (she speculated) was why it had not previously

Audrey Richards in middle age. A wonderful hostess with a spiky sense of humour, she entertained tricky guests by lighting matches with her toes.

attracted the attention of academics. Malinowski was impressed. Despite (as he confessed) having contributed to the current 'surfeit of sex' with four books on the topic, he wholeheartedly agreed with Richards about nutrition. Her approach proved that in 'the new science of man' it was still possible to be 'an explorer in untrodden fields'.

After the publication of *Hunger and Work in a Savage Tribe* (a very Malinowskian title), Richards signalled another small but significant break from her mentor with her decision never again to use the word 'savage' of the people with whom she worked.

Nutrition was a subject of interest in academic and civil service circles in the 1920s and 1930s to which Richards, with her undergraduate background in biology and her functionalist anthropological approach (as well as her skill with meringues: unusually for a woman of her background at this time, she was a famously 'wonderful' cook), was perfectly positioned to contribute. Diet and sustenance was a subject of significant interest in the developed as well as the developing world. In 1936, while Richards was in Zambia, John Boyd Orr published a report revealing that a fifth of all British children were nutritionally at risk, with half the children in lower-income groups underweight and nearly a quarter badly anaemic. Reformers like Boyd Orr and Richards hoped that nutritional understanding would improve the lives of people in the twentieth century, much as developments in sanitation had transformed conditions in the nineteenth.

When Richards returned to Zambia, Stewart Gore-Browne joked that she 'preferred porridge to sex'. Her second book, *Land, Labour and Diet in Northern Rhodesia* (1939), continued the nutritional theme, elaborating on the years of drought and plague of locusts she had witnessed during her on-and-off decade with the Bemba. Revealing 'the hard facts of survival and subsistence in rural areas of Africa drained of manpower by the developing mining and industrial towns', her work was placed firmly and specifically in the context of studying contemporary people in their current situation rather than focusing on 'primitive' or 'pre-literate' societies described in the traditional, vague 'ethnographic present'.

An article she wrote for *The Spectator* in 1935, responding to news of strikes and rioting among the copper miners of Rhodesia, presented a picture of floating, debt-plagued workers at the mercy of fluctuating copper prices and a colonial government raising taxes while cutting funds for 'native development' and education. The colonial government expected to be able to tax and exploit 'native' populations without reinvesting in their futures (or even bothering to learn their languages). For Richards, practical and down to earth, the predicament of the people with whom she worked was a problem she wanted to help solve.

Her dedication to scientific neutrality meant that she would not express strong views on policy or criticise officials, either colonial or, later, African, but as a 'do-gooder' (her phrase), she tried to formulate and shape research that would be 'helpful for "welfare and development" [the term then in use in the Colonial Development and Welfare Act]'. She wasn't naïve about what she hoped might be achieved, recognising later that 'many would deny the validity of our belief [that colonial government could be beneficial] ... especially those who feel that cultural and structural differences between the peoples inhabiting the ex-colonies should be obliterated as soon as possible'.

Disenchantment with colonialism from the 1930s onwards precipitated an existential crisis in anthropology, almost before it was truly established as a discipline. It was becoming increasingly difficult to justify the observation and analysis, however well intentioned, by European practitioners for a European audience of non-European 'subjects' dominated by European powers. By the 1970s, even the 'ideas and ideals of the Enlightenment', the intellectual inspiration for anthropology, were critiqued as being 'rooted in an unequal power encounter between the Western and Third World' by academics like Talal Asad. These ideas, as Asad observed, merely reinforced the gulf between Western elites, gathering power and information, and the people they studied, objectifying and demeaning those people.

Margery Perham, a historian with a special interest in British colonial administration, gave a talk on indirect rule at the Royal Society of Arts in London in March 1934. She argued that anthropologists, trained to understand 'native' societies in the round, could help improve the current 'tendency to misunderstand and the temptation to mishandle' so prevalent in colonial administration. Stella Thomas, a young Nigerian lawyer trained at Oxford, stood up to challenge her. Africans didn't want to be studied, she said, 'they wanted to be represented and to be given training so that they would be able to express themselves … At present the British were dictating to them, and the Africans had to do what they were told. They could not go against it because they had not got the power.' Perham responded afterwards that 'the new anthropologists' – perhaps she had Richards in mind – were 'no longer looking at Africans as specimens, nor seeking to preserve the past for the convenience of their researches'.

It was never an easy picture. While anthropology had the potential to aid colonial ambitions and ease colonial control by encouraging more 'enlightened' rule, it was more than capable of tainting colonial relations – on both sides. 'Natives' loathed being patronised and objectified; colonial officials resented the implied criticism of their policies of Westernisation, modernisation and industrialisation as well as the nature of their control. Not only that, eager to respect the differences between peoples, the anthropologist steeped in cultural relativity might almost unwittingly support customs she would condemn in her own society. 'At home, the anthropologist may be a natural subversive, a convinced opponent of traditional usage,' observed Claude Lévi-Strauss in 1961, 'but no sooner has he in focus a society different from his own than he becomes respectful of even the most conservative practices.'

Jomo Kenyatta, the first indigenous leader of Kenya, studied under Malinowski when Richards was a lecturer at LSE in the mid 1930s. Malinowski wrote the introduction to Kenyatta's influential 1938 book about the Kikuyu tribes to which he belonged, *Facing Mount Kenya*. By describing a self-sufficient indigenous group living orderly and virtuous lives, Kenyatta (who would go on in the 1960s to lead his country out of colonial rule) sought to use anthropology to challenge colonial control. He was so eager to demonstrate that his people were capable of independent government that he condoned female genital mutilation,* regarded by the Kikuyu as an essential rite of passage and bitterly opposed by most European missionaries and colonial officials.

Motivated by the desire to do good rather than to engage in theoretical debate, Richards stuck to her line, practical and positive. Her student, colleague and friend, Jean La Fontaine, said Richards simply 'saw herself as providing the knowledge needed by those in power to enable them to solve more effectively the problems which arose in the course of social change'. An anecdote she liked to tell, typically against herself, involved her first day at the Colonial Office during the Second World War, as related by La Fontaine: 'Having been introduced to her new job with instructions that stressed the necessity of taking action herself when possible, [Richards] was shown to her new office where she opened her first file, determined to follow her instructions to the letter and shirk no decision, however difficult ... [The] item proved to be a prison report that read "From the Governor of the Seychelles to the Colonial Office: I am pleased to report that the public conveniences installed last year continue to give satisfaction." After twenty minutes' agonised academic debate ...

---

* Female circumcision was an almost unthinkable concept for most Europeans. When in 1939 Elspeth Huxley, married to the son of Alfred Haddon's mentor, Thomas Huxley, was looking for a publisher for *Red Strangers*, her description of Kikuyu life at the time Europeans first arrived in Africa, Harold Macmillan agreed to publish it only on condition she remove a graphic description of female genital mutilation. Huxley went to Chatto & Windus instead, later commenting drily that 'our future Prime Minister couldn't take clitoridectomy'.

she took up her pen and wrote, in a firm hand, the single word "seen".'

After her time at the Colonial Office, helping to reorient research policy, Richards went back to Africa, this time to Uganda, between 1950 and 1956. As the first director of the newly formed Makerere Institute of Social Research in Kampala, she helped create, aided by her talent for informal and impromptu entertaining, 'a unique period of discourse between high government and intellectuals black and white'. Her party trick – lighting matches with her toes – broke the ice with everyone and her friendships were catholic, ranging from Ganda princes to the secretaries at the institute. Later she returned to Cambridge as director of the Africa Studies Centre and a fellow at Newnham. During her sixties and seventies, alongside Edmund Leach, she turned her ethnographic focus on Elmdon in Essex, where she lived, producing a genealogical study of the villagers and training young fieldworkers in the process. (In the 1930s, she had sent her students to the Rupert Street market near LSE to practise their field techniques.)

An article Richards wrote in 1974, at the age of seventy-five, casts some light on her attitude to her career as an unmarried female academic in the changing environment of the twentieth century, coloured by her functionalist training. The anthropologist's role, she wrote, is not to be shocked herself, but to note what shocks the women she is observing: 'I remember once sitting at the door of a tent set up in a Bemba village, consumed with pity at the fate of two or three women pounding with heavy pestles in large wooden mortars, with sweat pouring down their backs in the heat of the noon sun. But when they stopped to rest a moment one said to the other, "I don't know how these white women stand it! You would think her head would drop off bending over that book and writing, writing all the morning."'

Digging deeper into the rights and responsibilities women possess in different societies at different times and at different stages of their lives, she observed that ethnographic material seldom easily answers the many questions it raises: 'Is the woman in a more powerful position when she is in entire charge of an occupation necessary to her society and can exclude men from it, or when her work is completely interchangeable with that of men except during her pregnancy?

'The women's lib movement ... say that women are happy as housewives and mothers only because they have been brainwashed in childhood to believe that these are the roles appropriate for women and feel that these women should therefore be brainwashed in the opposite direction so that they become impatient of the mother and home role.' Like Margaret Mead, Richards wasn't convinced by this article of radical feminist faith; her approach, typically, was more common sense, less ideology: 'Our own predicament is surely that the modern family is so small and the housework is so much reduced by machine and convenience tools that housework is not enough to satisfy an active and intelligent woman although it is too great a burden to enable her to take another worthwhile job if she has small children.'

Still, times had changed since she had first arrived in Zambia. Malinowski had chosen a matrilineal society for her research because he believed it was appropriate for women to study women. 'When I got there,' she remembered with characteristic dry wit, 'you will not be surprised to find I found as many men as women!' In his tribute to her, her friend and colleague Raymond Firth said, 'she assumed rather than strove for equality with men'.

Throughout her career, her focus on practical or applied anthropology as opposed to pure or theoretical work does seem both to have been linked to her being a woman and a reason that she was never made a professor. Because she was concerned with how knowledge could usefully be applied, she was more interested in ethnographic detail, primary field data, than in theory; and, methodologically agnostic despite her Malinowskian functionalist training, she relished multidisciplinary collaboration. Both these tendencies, especially when combined with modest self-deprecation, hindered conventional academic promotion. Her clear, sensible anthropological thinking, though profound, appeared almost too simple – an unfashionable illusion among the densely written texts of the post-war academic world. Not until she was nearly sixty did she achieve the honour of becoming the first female president of the Royal Anthropological Institute; in 1967, she was the first woman anthropologist elected to the British Academy.

Richards' one regret in life was not having had children but she seems to have absorbed the unwritten rule of her family, caste and era that a woman could choose a family or a career, not both. Her two younger sisters had married fellows of All Souls, Oxford, and brought up families; her older sister, Gwynneth, who remained unmarried, was a social worker in Bermondsey. The satisfactions of Richards' career and deep friendships, attested to in a wealth of private letters and tributes after her death, were her consolation. Famously scatty, always leaving a trail of lost possessions in her wake, she was shrewd, affectionate and generous, her sparkling wit never malicious but marked by 'salt on her tongue'. Firth praised 'her sincerity of purpose and her commitment to scholarly and personal ideals' and her fresh attitude and outlook. 'Her lively, individual style of communication was shot through not so much with wit in the sense of a play upon words, as with her sense of humour expressed in a play upon elements in a social situation': even her laughter was somehow anthropological.

CHAPTER ELEVEN

The Trickster

Claude Lévi-Strauss in Brazil, 1938–1939

—●—

In the summer of 1935, while Audrey Richards was cycling from village to village among the Bemba and Franz Boas was considering his retirement from Columbia University, a young French ethnologist called Claude Lévi-Strauss visited his 'first' indigenous people in Brazil. He dreamed of walking in the footsteps of his anthropological hero, the Huguenot pastor and explorer Jean de Léry, who had written an account of his voyage to Brazil nearly 400 years earlier, and meeting people who had never before come into contact with Europeans – something not one of the other anthropologists in this book, from Boas in the Arctic onwards, could claim to have done.

On this first expedition, which took place during a holiday from his work as professor of sociology at the University of São Paulo, Lévi-Strauss accompanied an official from the Department for the Protection of Indians to the banks of the Rio Tibagy, where, in about 1914, after a century of colonial disruption and persecution, several groups of indigenous people had been moved on to reserves. Efforts had initially been made to integrate them into the modern world, for example encouraging them to settle in villages and live in houses, but they preferred to live nomadically, as they had always done, sleeping outside on the ground. Ignoring the cattle they'd been given for beef and milk, which they did not consume, the only 'civilised' things they kept were Western clothes and steel in the form of axes, knives and needles.

'So, to my great disappointment, the Tibagy Indians were neither completely "true Indians", nor, what was more important, "savages",' remembered Lévi-Strauss, who'd hoped for more 'poetry' from this encounter. He had been naïve. Instead, he found a people typical of many 'primitives' of the twentieth century, 'who had had civilisation brutally thrust upon them, but once the danger they were supposed to represent had been overcome no further interest had been taken in them'. Ancient traditions and borrowings from the developed world lay entangled on the pounded earth floors of their shelters, beautifully polished stone pestles beside cheap tin plates and salvaged petrol drums (no plastic – not yet). In a depressingly familiar pattern, like that observed by William Rivers in Melanesia decades earlier, Western diseases – malaria, tuberculosis and alcoholism – had devastated their numbers, while infant mortality rates had soared: an adult population of about 450 people on the São Jeronymo reserve had had 170 babies over the previous decade, but of those, 140 had died in infancy.

Lévi-Strauss was eager to find out about *koro*, a Tibagy treat of maggots that inhabited rotten tree trunks, but at first he could find no one willing to help him: 'Having been jeered at by the whites for eating these creatures, the Indians deny the charge and will not admit to liking them.' But despite their protestations, the forest was full of trees toppled by storms and hacked to pieces by *koro* hunters; 'and if you arrive unexpectedly at an Indian house, you may catch a glimpse of a bowl swarming with the treasured delicacy, before it is whisked out of sight'.

Only by insisting that he wanted to eat *koro* himself could Lévi-Strauss persuade 'a fever-stricken Indian' from the Kaingang tribe, alone in a deserted village, to take him into the jungle to find it. He pressed an axe into his guide's hand when they found a suitable tree; one blow revealed thousands of hollow chambers in the trunk, each containing a fat, pale grub. Watched by his impassive companion, Lévi-Strauss had no choice but to keep his word. Cautiously he decapitated the maggot and tasted the fatty goo that oozed out but there was no reason to hesitate: 'it had the consistency and delicacy of butter, and the flavour of coconut milk'.

This was Lévi-Strauss's initiation into fieldwork, a suitably epicurean moment for a man whose correspondence with his mother, a fabulous cook, is full of loving descriptions of food ('Artichoke hearts in mushroom sauce and seared fatty goose livers all pink inside – exquisite' read one letter from Bayonne in the early 1930s). For me, it reveals everything about Lévi-Strauss as

Claude Lévi-Strauss in a riverside camp kitchen in Brazil, 1938. His companion Lucinda (a young woolly monkey) can be glimpsed at his feet.

a fieldworker and a writer: his sense of having come too late to the study of people who had not followed the path of Western progress, his ability to see them on their own terms, with clarity but without sentimentality, and the elegiac, idiosyncratic beauty of his descriptions of their (almost) mutually experienced encounters.

Born in 1908, Lévi-Strauss was brought up in Paris. As a boy, he was a *flâneur* who loved walking the shimmering grey streets of the city and a collector, his first exotic treasure a Japanese stamp. His childhood was marked by three major forces: the primacy of art and music in his family, in which his grandfather was a composer and conductor and his father and two uncles were painters; the downward trajectory of his family's fortunes and social status, with intellectual achievement seen as the compensatory counterweight to this decline; and the experience of growing up as an assimilated Jew, albeit 'circumcised and bar mitzvahed', eating ham sandwiches in secret but discriminated against by non-Jewish children at school.

His path to anthropology was indirect. As he described it, he was a bored philosophy teacher when he read Robert Lowie's *Primitive Society* in the early 1930s; that, combined with his interests in geology and landscape, psychoanalysis and Marxism, made him receptive to an offer of teaching sociology as part of a French scheme at the newly founded University of São Paulo in 1934. Anthropology released him: 'My mind was able to escape from the claustrophobic, Turkish-bath atmosphere in which it was being imprisoned by the practice of philosophical reflection. Once it had got out into the open air, it felt refreshed and renewed. Like a city-dweller transported to the mountains, I became drunk with space, while my dazzled eyes measured the wealth and variety of the objects surrounding me.' Much later, he would refine this recollection, remarking that he never truly made 'a rupture with philosophy. Because in the end, we're always doing philosophy.'

Two years after his *koro*-tasting experience, Lévi-Strauss found himself in the wholesale shops around Réaumur–Sébastopol in Paris, anxiously surveying fish hooks, needles and glass beads in the preferred colours of white, black, red and yellow. Anything shoddy would inevitably be rejected by fastidious tribespeople on the expedition he planned. He had received the blessings of the most distinguished French social scientists, including Paul Rivet, director of the new Musée de l'Homme; written articles and obtained funding; and worked with the Brazilian authorities to assure them that any interesting objects of material culture he was able to procure would be divided between France and Brazil, in Brazil's favour.

Until the 1920s and 1930s, fieldwork was less a part of the anthropological tradition in continental Europe than in Britain and North America. The sociologist Émile Durkheim, who died in 1917, instituted a rigorous scientific approach to social phenomena, asking questions about what forces shaped societies and how they were held together; but after he became chair of education at the Sorbonne, he had little cause to leave the 5th arrondissement. In Britain, Alfred Radcliffe-Brown had been profoundly influenced by his scientific rigour and asceticism.

Durkheim's nephew and intellectual heir, Marcel Mauss, was another academic rooted in Paris (he held the chair of primitive religion at the École Pratique des Hautes Études and was later at the Collège de France), but his most important work, *The Gift* (1924), was explicitly anthropological, exploring the nature of reciprocal exchange across human history. Edward Evans-Pritchard,* in his

---

* Edward Evans-Pritchard was another great British anthropologist and fieldworker from this period, who studied under Charles Seligman and Bronislaw Malinowski at LSE in the late 1920s. His classic work, *Witchcraft, Oracles and Magic Among the Azande* (1937), reflects his research in north Central Africa. Between 1946 and 1970 he was Professor of Social Anthropology at Oxford.

1954 introduction to the first English translation, wrote, 'Mauss is telling us, quite pointedly ... how much we have lost, whatever else we may otherwise have gained, by the substitution of a rational economic system for a system in which an exchange of goods was not a mechanical but a moral transaction, bringing about and maintaining human, personal relationships between individuals and groups.' In his first lecture at the École Pratique des Hautes Études (EPHE), as early as 1901, Mauss had declared there were no 'uncivilised' people. His students were the first French social scientists to embrace fieldwork and Lévi-Strauss's eagerness to bring home from Brazil objects of material culture for the Musée de l'Homme owed much to Mauss's 'mysticism of the object'.

'Having no suspicion that the result would run counter to my plans,' Lévi-Strauss wrote later with devastating understatement, he envisaged a year's journey that would take him across the relatively unknown western part of the Brazilian plateau, the Mato Grosso (Thick Bush), from its state capital Cuiabá to the Rio Madeira. A Herculean effort by the army engineer Cândido Mariano da Silva Rondon had established a telegraph line in the early 1900s that linked the federal capital with the north-west frontier but, by 1938, made obsolete by short-wave radio, it had fallen into disrepair. Lévi-Strauss proposed to follow the length of this line.

Very few outsiders had travelled to this region. Rondon continued his work mapping Mato Grosso until 1919,* when politics overtook his career; he was the first outsider to make contact with the Nambikwara, one of the tribes Lévi-Strauss hoped to encounter, without being killed. With remarkable sangfroid, the British explorer (and possibly spy) Violet Cressy-Marcks had travelled through the Amazon and across the Andes to Peru in 1928–9, surveying part of the north-west Amazon basin, filming several remote tribes and writing a strangely pedestrian account of her journey.

No doubt Lévi-Strauss had heard the story of another British explorer, the geographer and archaeologist Percy Harrison Fawcett, who had set off from Cuiabá in April 1925 to find what he called the Lost City of Z,† which he believed was in the Mato Grosso. A friend of the novelists Arthur Conan Doyle and Rider Haggard, Fawcett travelled with just two companions, his twenty-one-year-old son Jack and a trusted friend of Jack's, Raleigh Rimell, alongside two Brazilian men, two horses, two dogs and eight mules – far more lightly than Lévi-Strauss fourteen years later. They had left instructions that no rescue expedition should be sent if they didn't return.

---

* Rondon, who was descended from several Brazilian indigenous groups including the Guaná and the Borono, was a lifetime champion of indigenous rights, first leader of the Indian Protection Service (whose motto, written by him, was 'Die if need be, never kill'), and helped establish the first Brazilian National Park for indigenous people. He was nominated for the Nobel Peace Prize in 1957, the year before he died, and the state of Rondônia, to the immediate west of Mato Grosso, is named for him.
† It is possible that the site of Kuhikugu, at the source of the Rio Xingu, uncovered by Michael Heckenberger in recent decades, is Fawcett's City of Z.

In January 1927, the Royal Geographical Society declared Fawcett, Jack and Rimell lost, presumed dead; their last surviving letter was dated just six weeks after their departure from Cuiabá, in which Fawcett told his wife that the three of them were setting off alone into unknown territory. Despite numerous theories – and several accounts of their murder by indigenous people, later proven to be false – no one has ever discovered what happened to them. Fawcett's story was one of several that demonstrated how easy it was to disappear in the wilds of the Brazilian bush; there was also talk of a group of Protestant missionaries killed in Juruena in 1930. If murderous tribespeople didn't loom out of the shadows with curare-tipped arrows, a traveller might easily die of starvation or exhaustion or any one of a variety of exotic diseases.

Although South America was the last major continent to attract the gaze of fieldworking anthropologists, by the 1930s a few of the more intrepid researchers had ventured there. A self-taught ethnologist, Curt Unkel, had lived among the Guaraní and Gê people, on and off, since his arrival from Germany in 1903, aged twenty; his indigenous friends gave him the name Nimuendajú ('the one who made himself a home'), by which he became known.* He corresponded with Robert Lowie, who translated some of his work, creating an academic link with North America. When Lévi-Strauss published his first

---

* Nimuendajú was killed by Tukúna tribespeople in 1945, allegedly for having sex with a girl who had not yet been initiated. All his archives were destroyed by a fire at the National Museum of Brazil in 2018.

Brazilian monograph in 1936, he sent it to Nimuendajú, who recommended it to Lowie: 'What is remarkable is that this man [Lévi-Strauss] ... who has only recently taken up anthropology, has so completely immersed himself in it in such a short time that he has managed to capture the sociological situation of the Bororo with great accuracy.' Three students from Columbia University were immersed in other tribes in the late 1930s, one of whom, Buell Quain, Lévi-Strauss befriended in Cuiabá in 1938.

More anxious, as he put it, to study America than human nature, Lévi-Strauss planned what his anthropological colleagues in England and the United States would have considered an old-fashioned survey rather than a period of intensive ethnographic immersion. He wanted to trace the whole of human history in the Americas by investigating the remaining indigenous people living in its most remote areas. Like an explorer from a bygone age,* he assembled a caravan made up of fifteen mules, thirty oxen (he likened them to duchesses, 'whose vapours, whims and fits of weariness had to be carefully studied') and an unreliable truck, as well as twenty local youths of Portuguese ancestry, hired at five francs a day, to be paid at the end of the expedition, with food and the use of a mule and a gun throughout it. Almost everywhere they went, once they left Cuiabá, their party would outnumber any group of indigenous people they met. The

---

* And rather belying the famous opening lines of *Tristes Tropiques*, the account he later wrote of this journey: 'I hate travelling and explorers ... Adventure has no place in the anthropologist's profession.'

Brazilian press noted that as well as hundreds of kilos of rice and sugar, they travelled with rifles, a typewriter, a radio set and a gramophone.

The academic team gathering in Cuiabá in May 1938 included Lévi-Strauss's wife of six years, Dina, also an anthropologist; a biologist, Jehan Albert Vellard, specialising in reptiles and spiders; and, at the last minute, a researcher at the National Museum in Rio, Luiz de Castro Faria, charged by its director with closely monitoring and reporting on the expedition's activities and, if necessary, ordering its immediate return. Faria was motivated by the desire to discover the anthropological roots of modern Brazil but ended up in his diary observing the ethnology of the expedition, while Lévi-Strauss, who called Faria 'Rio's eyes', had his gaze fixed on the distant past. Even before they set off, Dina was plagued by nightmares, confiding her fears to her diary: 'The adventure appears increasingly risky. Our "Nambiquara" really do have the worst reputation. Ferocious. Stories of terrible massacres. I am not prepared to die and I wish to return from this adventure. Otherwise, it will no longer be an adventure.' They set off at dawn on 6 June with Lévi-Strauss as leader on the most 'majestic animal' he had bought, 'a huge white mule'.

It would take several weeks for Lévi-Strauss to realise that, as he put it, 'the sense of time did not exist in the world I was now entering'. The scrubby country cut through by the derelict Rondon line was unlike anywhere else on earth. 'Completely virgin landscapes have a monotony which deprives their wildness of any significant value,' wrote Lévi-Strauss. 'They withhold themselves from man; instead of challenging him, they disintegrate under his gaze. But in this scrubland, which stretches endlessly into the distance, the incision of the *picada* [track along the telegraph line], the contorted silhouettes of the poles and the arcs of wire linking them to one another seem like incongruous objects floating in space, such as can be seen in Yves Tanguy's paintings. Being evidence of man's former presence and of the futility of his efforts, they mark the extreme limit he has tried to exceed, making it more obvious than it would have been without them.'

They covered the first five hundred kilometres in ten days, reaching the magnificent falls of Utiarity, on the far bank of which they saw two naked figures: the elusive Nambikwara. They pitched camp and set to work. At first it seemed that communication would be the biggest obstacle to their research – none of the twenty or so Nambikwara they met spoke any Portuguese and their language was 'completely inaccessible' – but alongside the usual wasps, mosquitoes and swarms of tiny blood-sucking midges, the biologist Vellard soon noticed a minuscule bee, nicknamed 'eye-licker', which infected the eyes of the humans it landed on. The infections rapidly progressed to purulent ophthalmia, 'gonorrhoeal in origin', among everyone there except Lévi-Strauss. By 22 July, Dina, the worst affected, had to abandon the expedition. Lévi-Strauss accompanied her to Cuiabá, from where she travelled slowly home to Paris, and then returned to the Nambikwara, with whom he stayed for a further two months; their marriage never recovered and Dina would barely merit a mention in his 1955 account of this period, *Tristes Tropiques*.

The Nambikwara were very different from the tribes Lévi-Strauss had previously encountered during his time in Brazil. Some, like the reservation tribes, were tragically tainted by their contact with Europeans, the human equivalent of the devastated landscapes he described covered with felled stumps, abandoned after being briefly used for farming or mining and condemned to poverty and decay. Others, like the Bororo, were remote enough (or lived in areas undesirable enough) to have retained their complex kinship arrangements, deliberately laid-out circular villages, clan myths, festivals, 'simple and constantly repeated' ritual songs, highly prized accessories – lip plugs, necklaces, headdresses made of jaguar teeth, toucan beaks, macaw tail feathers and egret tufts – and beautiful face and body painting traditions, as with the Caduveo, who also produced ornamented pottery and painted hides.

But the Nambikwara had nothing. Everything they owned fitted into the woven cane baskets their women carried from camp to camp: mostly raw materials like wood for kindling fires, lumps of wax or resin, shells, feathers and porcupine quills alongside fruit, palm-wood bows, nut husks and poisonous insects. Women might wear a thin string of shell beads tied around their waist,

Nambikwara people in a straw shelter. One of the woven cane baskets used by the women to carry the tribe's few possessions is just inside the shelter on the left.

while men scorned even the straw penis sheath of other tribes, opting instead for 'a straw tassel hanging from a belt above the sexual organs'. Lévi-Strauss had found 'one of the most rudimentary forms of social and political organisation that could possibly be imagined … [and] it was precisely that experience which was difficult to grasp. I had been looking for a society reduced to its simplest expression. That of the Nambikwara was so truly simple that all I could find in it was individual human beings.'

Today Lévi-Strauss's field notebooks are in the Bibliothèque Nationale de France. Colour charts, musical transcriptions and existential musings lie jumbled alongside kinship diagrams, lists of words, taxonomic groups and linguistic notes. He describes childbirth and defecation, the weather, the landscape. His sketches depict a variety of monkeys and other fauna, string games, penis sheaths, basket-weaving patterns, intricate facial decoration, and maps. Taken together, they demonstrate the detail and breadth of his ethnographic technique and imagination as well as his grounding in art and music. As Alfred Haddon had observed fifty years earlier, the ideal anthropologist would possess a multiplicity of talents.

By the autumn of 1938, Lévi-Strauss had been with the Nambikwara since June, travelling beyond the savannah to the southern edge of the Amazon. 'Adventure had been watered down into boredom', despite his constant companion, Lucinda, a young woolly monkey given to him by a Nambikwara woman. She clung awkwardly to his left ankle as he walked by day and slept a whiskey- and condensed-milk-soaked sleep by night. In these green fields surrounded by rich damp forest, full of game, Lévi-Strauss noted new gastronomic experiments: skewered hummingbirds or parrots, roasted and flambéed with whiskey, and grilled caiman tail. For the first time in ages, he and his companions took off their pith helmets, overalls and knee boots to wash properly: the playful Nambikwara women, with their propensity to steal the soap, had earlier made washing tricky.

(Left) *Mischievous women and children threatening to steal Lévi-Strauss's soap.*

(Right) *Treacherous driving conditions in the Mato Grosso of Brazil.*

At last, Lévi-Strauss came face to face with a tribe 'whom no white man had ever seen before and who might never be seen again': the Mundé (more properly called the Aikanã), entirely charming but 'Alas! they were only too savage.' He could look at them, touch them, 'but I could not understand them. I had been given, at one and the same time, my reward and my punishment. Was it not my mistake, and the mistake of my profession, to believe that men are always men? that some are more deserving of interest and attention because they astonish us by the colour of their skin and their customs? I had only to succeed in guessing what they were like for them to be deprived of their strangeness: in which case, I might just as well have stayed in my village. Or if, as was the case here, they retained their strangeness, I could make no use of it, since I was incapable of even grasping what it consisted of.'

Although he was immersed in conditions about which, as an ethnologist, he had long dreamed, Lévi-Strauss was never reconciled with the difficulties of fieldwork, perhaps at their most challenging here: 'You have to be up at daybreak, and then remain awake until the last native has gone to sleep, and even sometimes watch over him as he sleeps; you have to try to make yourself inconspicuous, while being constantly present; see everything, remember everything, note everything; display an embarrassing degree of indiscretion, coax information out of a snotty-nosed urchin and be ready to make the most of a moment's obligingness or carelessness; or alternatively, for days on end you have to repress all curiosity and withdraw into an attitude of reserve because of some sudden change of mood on behalf of the tribe. As he practises his profession, the anthropologist is consumed by doubts: has he really abandoned his native setting, his friends and his way of life, spent such considerable amounts of money and energy, and endangered his health, for the sole purpose of making his presence acceptable to a score of miserable creatures doomed to early extinction, whose chief occupations meanwhile are delousing themselves and sleeping, and on whose whims the success or failure of his mission depends?' He found himself in a cycle of depression and uncertainty, with Chopin's Étude

Op. 10, No. 3, the 'Tristesse', repeating in his head. It was a piece of which he was not fond.

By December, he was back in Cuiabá and, three months later, in March 1939, less than a year after he had set off on his expedition, he found himself back in Paris, with 'three trunks, six cases and a canvas parcel' containing the 745 objects destined for the Musée de l'Homme (760 had been left in Brazil, but they have since been lost). Only his monkey, Lucinda, remained to remind him of the hallucinatory experiences of the previous year. He was thirty and Europe was on the brink of war.

He spent the first few months of his return in the back rooms of the Musée de l'Homme, supervised by Paul Rivet as he catalogued his Brazilian objects, saturated in protective creosote, preparatory to writing up his dissertation. But then events overtook him: although he joined the French army, when France surrendered to Germany even secular Jews were effectively denaturalised. In 1941, having come to terms at last with the gravity of his position in France, he managed to escape to the United States via Martinique and Puerto Rico – this episode forms the opening to *Tristes Tropiques* – where the New School of Social Research in New York had a teaching role for him. Robert Lowie, who had heard about Lévi-Strauss's Brazilian work from Nimuendajú, acted as referee, part of a concerted Boasian campaign to bring European Jewish academics to safety in the United States.

In New York he found himself at the heart of a group of European exiles that included surrealist artists and writers like Albert Camus. He personified the link between folklore, primitivism and modernism that didn't really exist in anglophone anthropology: in France, and particularly in Lévi-Strauss, ethnology combined a modern faith in science with an avant-garde aesthetic fascination with the exotic and the unusual. Accompanied by the surrealists André Breton and Max Ernst, fellow exiles, he roamed Manhattan's dusty antique shops, hunting down Amerindian artefacts, fortified by the exotic foods of the city: 'barley meal fried chicken' in Harlem, 'Panamanian turtle eggs, moose stew, Syrian kefirs, softshell crab, oyster soup, Mexican palm tree worms' or the delights of Chinatown. After the war, he extended his stay several years to act as part-time cultural attaché to the French embassy.

During these years, Lévi-Strauss continued his anthropological work, teaching at the New School (in rapidly improving English), writing articles and researching his dissertation. He dined separately with Ruth Benedict and Ralph Linton, Boas's unpopular replacement as head of Columbia's anthropology department, and listened to them railing against one another; he met Margaret Mead and continued his working friendships with Lowie (who helped edit his thesis) and Alfred Kroeber, when they were in New York from California. Alfred Métraux,* a Swiss ethnologist of South America, was at Yale and became a close friend; and Lévi-Strauss was fundamentally influenced by his friendship with the Russian structural linguist Roman Jakobson, who introduced him to the work of Ferdinand de Saussure.

---

* His ex-wife, Rhoda, would become the companion for many years of Margaret Mead.

In December 1942, he was invited to the Columbia Faculty Club for a lunch for Paul Rivet, who had escaped Vichy Paris and was en route for Columbia, where he would establish the Anthropological Institute and Museum. Franz Boas, aged eighty-four, appeared in an ancient fur hat, evidently dating back to his Inuit expedition sixty years earlier. After a festive lunch surrounded by friends and colleagues, he rose to remind them of the need to remain vigilant in the struggle against racial prejudice before falling back into his chair, where he was caught by Lévi-Strauss, sitting beside him. Rivet, who had started his career as an army doctor, vainly tried to revive him but Boas was dead. The symbolism of this moment was lost on no one, not least Lévi-Strauss, who saw himself as heir to and uniting the traditions of his heroes, Mauss in Europe and Boas in the United States.

Late in 1947, Lévi-Strauss returned to France. At the Sorbonne in June 1948, he presented his doctoral theses, the major part of which would become *The Elemental Structures of Kinship* on its publication the following year. His minor thesis, 'The Family and Social Life of the Nambikwara', was more explicitly based on his Brazilian expedition, and as such was simpler for the committee to assess. *The Elemental Structures of Kinship* was global in its scope and deliberately theoretical, for example in its analysis of the relationship between kinship systems and language, marking a distinct move from the fieldwork monograph model that had become traditional in the discipline for theses over the previous half-century. Dismissive of the so-called discordant trumpeters of contemporary anthropological functionalism, Malinowski and Radcliffe-Brown, Lévi-Strauss made a return to the philosophical sweep and scope of Edvard Westermarck (whose *History of Human Marriage* had been one of the four books on the reading list of his sociology course at São Paulo in 1935)* while producing a work that felt entirely modern. He held that the incest taboo, on which Westermarck had worked decades earlier, was the very foundation of society, since it impelled men to look outside their families for a mate; he called this the alliance theory. Impressed but disconcerted, for five hours his examiners debated this 'monumental ... work of a philosopher', dense with algebraic formulae and written with the opacity that was to become the hallmark of Lévi-Strauss's academic style, before awarding him 'high honours'.

This success was closely followed by several years of professional setbacks and personal upheaval, during which his father died and he remarried and quickly divorced. He was rejected from prestigious roles at the Sorbonne and the Collège de France and his work at EPHE, UNESCO (producing the influential *Race and History* in 1952) and the Institut d'ethnologie of the Musée de l'Homme was slow to gather momentum. In 1954, the year he married his third (and last) wife, Monique Roman, frustratedly, furiously he embarked on what would become *Tristes Tropiques*.

---

* The others were Durkheim's *Elementary Forms of Religious Life* (to which the title of the course, 'Elementary Forms of Social Life' referred), Lowie's *Primitive Society*, and Arnold van Gennep's *L'état actuel du problème totémique*.

Conceived in the shadow of the Second World War, coloured by what he had experienced himself and witnessed of mankind, *Tristes Tropiques* was melancholy and elegiac, mourning the loss of the richness and beauty of vanishing cultures the modern world could not understand and was condemned to destroy. Human history was less a matter of progress than of decline. 'I lose on both counts,' he wrote of his UN-sponsored travels in crumbling Lahore. 'All that I perceive offends me, and I constantly reproach myself for not seeing as much as I should.' It is a curious book, unclassifiable; anthropological, certainly, but more a philosophical musing on anthropology than a book of anthropology. Written as a memoir, it is steeped in the existentialism of Lévi-Strauss's Parisian contemporaries (even though he often disagreed with them) and recalls other great twentieth-century works of deracination, such as Vladimir Nabokov's *Speak, Memory* (1951). At last the anthropologist turns his gaze upon himself: prefigured by Malinowski, Benedict and Hurston, Lévi-Strauss draws aside the veil and steps into the light, seeking to describe 'those feverish moments when, notebook in hand, I jotted down second by second the expressions which would perhaps enable me to fix those evanescent and ever renewed forms'.

Anthropologists viewed it with professional detachment. Taken as a record of ethnographic practice, drily observed Paul Rabinow (part of a later generation of fieldworkers), it was 'an overcompensation for the author's shortcomings in the bush'. These reservations were not felt by its general readers, who received it rapturously. The judges of the Goncourt Prize declared that they wished they could have awarded Lévi-Strauss the trophy, if only *Tristes Tropiques* had been a novel; Susan Sontag, in *The New York Review of Books*, compared it to Montaigne and Freud, calling it 'one of the great books of our century ... an intellectual autobiography, an exemplary personal history in which a whole view of the human situation, an entire sensibility is elaborated'.

From this point onwards, despite the admitted inconsistency of his fieldwork (on which he did not embark again), Lévi-Strauss's position at the heart of French – indeed, global – intellectual life was secure. In 1959, aged fifty-one, he was elected to the chair of social anthropology at the Collège de France, a position he held until 1982. Challenging, ambitious volumes of anthropological work followed, marked by the playful but dazzlingly complex thinking that would confound students of anthropology for decades to come and redefine the intellectual arena of the twentieth century. His structuralist manifesto, *Structural Anthropology*, came out in 1958, with volume two appearing in 1973; *La Pensée Sauvage* in 1962; *Totemism* in 1963; and *Mythologies* over four volumes between 1964 and 1971.

One of the cultural archetypes Lévi-Strauss examined was the trickster, a disruptive, ambiguous character who features in almost

---

* Thankfully a biographical study of anthropologists in the field before the Second World War is not the place to discuss structuralism and post-structuralism, the academic movements to which Lévi-Strauss's anthropological work was one of the seminal contributors; suffice it to say they dominated intellectual life from the 1960s to the 1980s and beyond. His work can be thought of as reassuringly difficult.

every known human mythology. Dual in nature, he contains a combination of opposite qualities, often mediating between them. This is how I see Lévi-Strauss himself, who used anthropology to question the traditions and achievements of Western society while extolling (and arguably romanticising) the virtues of non-Western civilisations. 'Anthropology reflects, on the epistemological level, a state of affairs in which one part of mankind treats the other as an object,' he wrote. 'Primitive mentality' was a modern fiction, invented to justify to European students of other cultures what they believed was their own inalienable superiority: rather, he argued, 'savage' and 'civilised' minds were essentially the same.

Lévi-Strauss's suspicion that the Western world did not have all the answers, formulated in his years in Brazil in the 1930s, only grew stronger as the decades of his life and career unfolded. In this, as in his passionate environmentalism and love for animals (he became a vegetarian in his sixties), he would grow ever more relevant to subsequent generations. He described in *Tristes Tropiques* the early encounters between Amerindians and European colonists in the sixteenth century. The Spanish observed the behaviour of the indigenous inhabitants of Hispaniola (Haiti and Santo Domingo today), their hairlessness, nakedness, cannibalism and fondness for eating insects, and dismissed them as animals, nothing more than potential slaves; the islanders of Puerto Rico meanwhile systematically captured white men and drowned them, keeping watch over the corpses for several weeks to see whether they decayed: 'Two conclusions emerge from a comparison between these different processes of investigation: the whites trusted to social science, whereas the Indians had confidence in natural science; and while the whites maintained that the Indians were beasts, the Indians did no more than suspect the whites might be gods. Both attitudes show equal ignorance, but the Indians' behaviour certainly had greater human dignity.'

One of the closing chapters of *Tristes Tropiques* contains a meditation on the different methods of rum production Lévi-Strauss had observed in the Caribbean. In Puerto Rico, where rum manufacture and export was big business, it was distilled in gleaming modern enamel and chromium factories; in Martinique, the ancient wooden vats were encrusted with waste. Martinique rum, though, was mellow, scented and subtle compared to the harsh coarseness of Puerto Rican rum. 'To me, this contrast illustrates the paradox of civilisation,' he wrote. 'Its charms are due essentially to the various residues it carries along with it, although this does not absolve us of the obligation to purify the stream.'

Perhaps uniquely among this group of anthropologists, Claude Lévi-Strauss recognised that anthropology, once it had accepted the value of different societies and individuals, had no real answer for the challenges that faced people and peoples as the march of history progressed. 'Civilisation' impoverished humanity as much as it enriched it; anthropology might just as easily be termed entropology. All one could hope to do was 'to spread humanism to all humanity'. This was the lesson he had learned in his months with the Nambikwara, 'savages' who had so little and yet understood so much. They had touched him deeply, revealing to him 'an immense kindness, a profoundly carefree attitude, a naïve and charming animal satisfaction and – binding these various feelings together – something which might be called the most truthful and moving expression of human love'.

Conclusion

———•———

With the advent of the Second World War, this formative period in anthropology's history drew to a close. British and American anthropologists dedicated themselves afresh to refuting the racism that was so pernicious a motivation for German aggression and expansionism, both during the war and afterwards, often working with the newly formed United Nations. On 10 December 1948, the UN released its Universal Declaration of Human Rights, intended to prevent the atrocities of the war from ever being committed again. Although its creators were predominantly lawyers, led by the redoubtable figure of Eleanor Roosevelt, they were fundamentally influenced by the work done in anthropology over the preceding decades. As one of their number, the Chilean jurist and diplomat Hernán Santa Cruz, wrote, 'I perceived clearly that I was participating in a truly significant historic event in which a consensus had been reached as to the supreme value of the human person, a value that did not originate in the decision of a worldly power, but rather in the fact of existing.' The human person: no mention of race or colour, not a savage, not a subject, not even a woman or a man. A designation of utter simplicity, but one that would have been unthinkable fifty years earlier.

There followed decades during which anthropology retreated from its previous engagement with wider society, becoming (as George Stocking would put it) 'the most highly academicised of the social sciences' – which is saying something. As the ideas, especially linguistic, of Claude Lévi-Strauss filtered through the world of academia, they had a profound influence on other disciplines and thinkers, notably the psychoanalyst Jacques Lacan and the philosophers Jacques Derrida and Michel Foucault, whose insights about knowledge and power resonated especially deeply in anthropology. At the same time, the methods and focus of the discipline were becoming extraordinarily broad and varied, moving away from the study of 'primitive' people to encompassing anything humans do or have done. In 1986, the anthropologist James Peacock suggested viewing anthropology as an orchestra, with each of its multiplicity of specialities representing an instrument playing in concert with all the others.

In parallel with the post-war collapse of the colonial world, fieldwork for a time was regarded, often by anthropologists themselves, with at best ambivalence and cynicism. Conquest and repression were seen as the unavoidable consequences of the spread of Western civilisation, with those who had taken (in the words of Jean and John Comaroff in 1992) 'to the field to subvert Western universalisms with non-Western particularities ... [standing] accused of having served the cause of imperialism'. How were fieldworkers to rinse themselves clean of ethnocentricity when their very ethnocentricity was part of what had made them curious about the wider world in the first place? This was the challenge that faced anthropology as it approached the end of its first century as an academic discipline.

One by one, the individual reliability of influential fieldworkers like Bronislaw Malinowski, Ruth Benedict and Margaret Mead was called into question – as if they had been claiming their work was scientific fact, rather than interpretation. In the 1970s, Raoul Naroll demonstrated that ethnographers who remained in the field for a long time with a single group were more likely to report beliefs in witchcraft: was this because people trusted them more and were therefore more likely to confide difficult truths, or had the anthropologists somehow been influenced by their extended time in the field? 'Different ethnographers sometimes report different "realities",' explained Robert Rubinstein. 'Moreover issues of power and perspective, questions of how authoritative knowledge is legitimated, of self-awareness and authenticity of voice in the presentation of data, and of the constraints of the historical and cultural contexts within which knowledge develops complicate the description and understanding of cultural and social life.'

In *The Interpretation of Cultures* (1973), Clifford Geertz, a compelling interrogator of anthropology from within, put a new spin on the concept of culture as articulated by Boas and his students: 'Believing, with Max Weber, that man is an animal suspended in webs of significance he himself has spun, I take culture to be those webs, and the analysis of it to be therefore not an experimental science in search of law but an interpretative one in search of meaning.' Gradually anthropology was becoming more open to ambiguity, the sense (as the Comaroffs put it) that 'the meaningful world is always fluid'. 'Culture is interpretation,' Paul Rabinow concluded in *Reflections on Fieldwork in Morocco* (1977): 'There is no privileged position, no absolute perspective, and no valid way to eliminate consciousness from our activities or those of others.'

The irony, continued Rabinow, is that our society, Western civilisation and indeed anthropology 'pretends to itself that it is investing them [primitive peoples] with nobility at the very time when it is completing their destruction ... Not content with having eliminated savage life, and unaware even of having done so, it feels the need feverishly to appease the nostalgic cannibalism of history with the shadows of those that history has already destroyed.' Indigenous people who had helped the researchers who arrived uninvited in their homes suffered from no such illusions. One of Radcliffe-Brown's colleagues at the University of Chicago, Dorothy Eggan, was working with Hopi Indians at the start of World War II. 'I pity you and I don't envy you,' an elder told her. 'You have more goods than we have, but you don't have peace ever. It is better to die in famine than in war.'

Conclusion

From the perspective of 2022, the anthropologists outlined here seem almost more old-fashioned in their anthropocentrism than anything else. The biologist Merlin Sheldrake, fascinated by fungi and the way (for example) slime mould can out-think humans in certain situations, calls it species narcissism; Gaia theorists and specialists in artificial intelligence might take a similar view. If the study of man provided the intellectual background for the twentieth century, is it too much to hope that understanding and looking after the earth as a whole might be the theme of the twenty-first? Margaret Mead's focus was more on people than the planet, but she might well have agreed: 'Looking back, it seems to me that the most significant change in these years [the 1960s] has been our growing awareness of the interconnectedness of problems and the interdependence of people not only in our own country but everywhere on our crowded planet.' We need to work together to solve the problems that confront us and, despite its complications, fieldwork, which requires us to walk a few steps in someone else's shoes, can still help us achieve that.

Perhaps what anthropology needed to address was the fact that Western civilisation, the initial source of its enquiry, is neither a completely beneficial influence, as nineteenth-century anthropologists once believed, something to be aspired towards, nor a one-dimensional force of iniquity that taints everything with which it comes into contact. Like these individual anthropologists, it is complex, at once disturbing and rewarding, eluding attempts at classification. Building on Lévi-Strauss's ambiguity about civilisation, and by extension his own work, I see it as a powerful drug that seeps irretrievably into everything it touches and can be positive or negative depending on the context. Climate change, the devastation we have wrought as a species in our rush to be supreme and the urgency of our need to change our behaviour are the most obvious challenges we contemplate at the moment, but there are plenty of other examples. Chemotherapy kills healthy cells as well as cancerous ones; fast food and microwave meals are cheap and convenient but nutritionless and devastating to nature; the internet can provide support for lonely people or make them feel even more isolated. Before we learned to medicalise death, more people died peacefully of old age at home. Before electrically powered radios and television, people made music, sang and told stories together. What we can lose when we become 'civilised' is as important and as valuable as what we might gain. These early fieldworkers, culminating with Lévi-Strauss, taught us that we must honour and preserve the things that make us 'savage' even while we strive for the things that make us 'civilised', because it is the combination of both that makes us human.

Acknowledgements

——•——

Hearty thanks are due first of all to Kirty Topiwala, who not only suggested the idea for this book to me but also gave her time with thoughtfulness and generosity while I grappled with how it might be done. I was lucky too that at the start of this project, as a Royal Literary Fund Fellow, I had access to the University of London's magnificent library at Senate House; the privilege of freely roaming the anthropology shelves there was hugely useful.

Many thanks to Storm Woodward and the newly formed Anthropology and Psychology Society of Monmouth School for inviting me to speak to them.

James Nightingale, Will Atkinson, Karen Duffy, Hana Qureshi and all at Atlantic have been, once again, the most supportive of publishers: thank you all. I am very grateful to Jane Selley for her scrupulous copy-editing (all mistakes that remain, errant commas included, are my own) and to Alex Bell for the index. And as ever, thank you, dear Clare, for your wise and encouraging counsel.

Acknowledgements

I would like to dedicate this book to the Miller family, Helen and Irving, Josh and Abby, who gave me such a loving welcome when, as a know-it-all six-year-old, I reluctantly entered their lives. Their immensely impressive careers in social work, the coal face of the social sciences, both as practitioners and academics, have been an inspiration to me, as well as introducing me early on to anthropological thinkers, theories and concepts that I have been fascinated to encounter again in researching this book.

Finally, my love to numbers one through four. You know where you stand.

Notes and Bibliography

Each chapter has a slightly different character and focus, dictated by the variety of material available. By necessity in a survey of this kind, I've depended, very gratefully, on the work of biographers and scholars of anthropology, as well as original sources. I've provided a list of my most important sources for each chapter, along with the more general bibliography that follows. Books are mentioned only once, so for example Emmanuelle Loyer's biography of Lévi-Strauss appears in the Introduction rather than the chapter on Lévi-Strauss; where the place of publication isn't given, it is London; and if I haven't used the first published edition, I've indicated in parentheses the year the book was first published.

I heartily recommend Umberto Eco's satirical essay 'Industry and Sexual Repression in a Po Valley Society' (1962), which could accompany many of the chapters above but was inspired by Bronislaw Malinowski. A 2017 BBC Radio 4 series (now available as a podcast), *From Savage to Self*, written and presented by Dr Farrah Jarral, is a great introduction to the history of anthropology.

Introduction

Banks, J., *Endeavour Journal*, available online

Dewalt, K. M. and B. R., *Participant Observation: A Guide for Fieldworkers*, Oxford, 2002

Frazer, J., *The Golden Bough*, Oxford, 1994 (1890)

Helm, J., ed., *Pioneers of American Anthropology: The Uses of Biography*, Seattle, 1966 (especially the chapter by Jacob Gruber)

Holmes, R., *The Age of Wonder: How the Romantic Generation Discovered the Beauty and Terror of Science*, 2009

Hyatt, M., *Franz Boas, Social Activist: The Dynamics of Ethnicity*, Westport, 1990

Loyer, E., *Lévi-Strauss: A Biography*, Cambridge, 2019 (quoting Viktor Karady on innovative scientific fields offering unconventional paths to success)

Mead, M., *Coming of Age in Samoa*, 1953 (1928)

O'Brien, P., *Joseph Banks: A, Life*, 1987

Peacock, J. L., *The Anthropological Lens*, 1986

Rohner, R. P., *The Ethnography of Franz Boas*, Chicago, 1969

Stocking, G. W., *Race, Culture and Evolution: Essays in the History of Anthropology*, Chicago, 1982

Vermeulen, H. F., *Before Boas: The Genesis of Ethnography and Ethnology in the German Enlightenment*, Lincoln, 2015

Chapters 1 and 6: Boas

Adams, W. Y., *The Boasians: Founding Fathers and Mothers of American Anthropology*, Lanham, 2016

Anshen, R. D., *Freedom: Its Meaning*, 1942

Boas, F., *Anthropology and Modern Life*, 1929

——*The Central Eskimo*, Toronto, 1974 (1888)

——*The Mind of Primitive Man*, New York, 1929 (1911)

——*Race, Language and Culture*, New York, 1940

——*Primitive Art*, New York, 1955 (1927)

——*Handbook of American Indian Languages*, Washington, 1911

——*Changes in Bodily Form of Descendants of Immigrants*, New York, 2006 (1912)

Boas, F., ed., *General Anthropology*, New York, 1938

Boas, F., et al., *Anthropology in North America*, New York, 1915 (facsimile 1981)

Bouchard, R., and D. Kennedy, trans. and ed., *Indian Myths and Legends from the North Pacific Coast of America* (trans. of Boas's *Indianische Sagen*), Vancouver, 2002

Cole, D., *Franz Boas: The Early Years 1858–1906*, Seattle, 1999

——'The Value of a Person', in Stocking, *History*, Vol. 1

Darnell, R., ed., *The Franz Boas Papers, Volume 1*, Lincoln, 2015

——*And Along Came Boas: Continuity and Revolution in Americanist Anthropology*, Amsterdam, 1998

Fabian, A., *The Skull Collectors: Race, Science, and America's Unburied Dead*, Chicago, 2010

Herskovits, M., *Franz Boas: The Science of Man in the Making*, New York, 1953

Jonaitis, A., ed., *A Wealth of Thought: Franz Boas on Native American Art*, Seattle, 1995

King, C., *Gods of the Upper Air: How a Circle of Renegade Anthropologists Reinvented Race, Sex and Gender in the Twentieth Century*, 2019

Laufer, B., ed., *Boas Anniversary Volume: Anthropological Papers Written in Honour of Franz Boas*, New York, 1906

Lowie, R., *Robert H. Lowie, Ethnologist*, Berkeley, 1959

MacDonald, K., *The Culture of Critique: An Evolutionary Analysis of Jewish Involvement in Twentieth-Century Intellectual and Political Movements*, Westport, 1998

Müller-Wille, L., *The Franz Boas Enigma: Inuit, Arctic, and Sciences*, Montreal, 2014

Müller-Wille, L., ed., *Franz Boas Among the Inuit of Baffin Island, 1883–1884*, Toronto, 1998

Murphy, R. F., *Robert H. Lowie*, 1972

Ovington, M. W., *Half a Man: The Status of the Negro in New York*, New York, 1911

Stocking, G. W., ed., *The Shaping of American Anthropology, 1883–1911: A Franz Boas Reader*, New York, 1974

——*The Ethnographer's Magic and Other Essays in the History of Anthropology*, Madison, 1992

——*History of Anthropology*, Vols. 1–6, Madison, 1989

White, L., in *Monographs in Cultural Anthropology* 52 (4), 1966

Williams, V. J., *Rethinking Race: Franz Boas and his Contemporaries*, Lexington, 1996

Zumwalt, R. L., and W. S. Willis, *Franz Boas and W. E. B. Du Bois at Atlanta University, 1906*, Philadelphia, 2008

Charles King's *Gods of the Upper Air* came out while I was writing this book. With its focus on Boas and his female students, particularly Margaret Mead, it contains fantastic archival detail.

Chapter 2: Haddon and Rivers

Aldis, H. G., ed., *The Cambridge University Library*, 1922 (see Rivers' essay, 'History and Ethnology')

Barker, P., the Regeneration Trilogy, W. 1991–5

Bartlett, F. C., 'William Halse Rivers Rivers 1864–1922', *American Journal of Psychology* 34 (2), 1934

Bennett, A., *Things That Have Interested Me*, Vol. 2, 1923

Haddon, A., 'William Halse Rivers Rivers 1864–1922', *Nature* 109, 17 June 1922

Hviding, E., and C. Berg, *The Ethnographic Experiment: A. M. Hocart and W. H. R. Rivers in Island Melanesia, 1908*, New York, 2014

Kuper, A., *The Invention of Primitive Society: Transformations of an Illusion*, 1988

Langham, I., *The Building of British Social Anthropology: W. H. R. Rivers and his Disciples in the Development of Kinship Studies*, 1981

Leach, E., entry on Rivers in the *International Encyclopaedia of the Social Sciences*, Vol. 13, 1968

Quiggin, A., *Haddon the Head Hunter*, Cambridge, 1942

Rivers, W. H. R., *The Todas*, 1906

——*Kinship and Social Organisation*, 1914

——*The History of Melanesian Society*, Cambridge, 1914

——*Instinct and the Unconscious*, Cambridge, 1920

——*Psychology and Ethnology*, 1926

——'Essay on Anthropology', *Notes & Queries*, 4th edn, 1912

Rivers, W. H. R., A. R. Jenks and S. G. Morley, *Reports upon the Present Condition and Future Needs of the Science of Anthropology*, Washington, 1913

Sassoon, S., *The Complete Memoirs of George Sherston*, 1983

Slobodin, R., *Rivers*, Stroud, 1997 (1978)

Stocking, G., *After Tylor: British Social Anthropology 1888–1951*, Madison, 1995

Chapter 3: Westermarck

Grove, G., *71 Days Camping in Morocco*, 1902

Lyons, A. P. and H. D., *Sexualities in Anthropology: A Reader*, Chichester, 2011

Rabinow, P., *Reflections on Fieldwork*, Berkeley, 1977

Shankland, D., ed., *Westermarck*, Canon Pyon, 2014

Stroup, T., *Edward Westermarck: Essays on his Life and Works*, Helsinki, 1982

Suolina, K., C. af Hällström and T. Lahtinen, *Portraying Morocco: Edward Westermarck's Fieldwork and Photographs 1898–1913*, Åbo, 2000

Thomson, J., *Travels in the Atlas and Southern Morocco*, 1889

Westermarck, E., *The History of Human Marriage*, 1901

——*Memories of my Life*, trans. A. Barwell, 1929

——*Marriage Ceremonies in Morocco*, 1914

——*Ritual and Belief in Morocco*, 1926

——*Wit and Wisdom in Morocco*, 1930

——*The Origin and Development of Moral Ideas*, 1906

Wharton, E., *In Morocco*, 1920

See also Westermarck's photo archive on the website of Åbo Akedemi.

Chapter 4: Bates and Radcliffe-Brown

Bates, D., 'Notes on the Topography of the Northern Portion of Western Australia', Royal Geographical Society of Australia, 1905–6

Blackburn, J., *Daisy Bates in the Desert*, 1997 (1994)

Grant Watson, E. L., *But To What Purpose?*, 1946

Leach, E., 'Social Anthropology: A Natural Science of Society?', The Radcliffe-Brown Lecture at the British Academy, 20 May 1976

Powdermaker, H., *Stranger and Friend: The Way of an Anthropologist*, 1967

Radcliffe-Brown, A., *The Andaman Islanders: A Study in Social Anthropology*, Cambridge, 1922

——*The Social Organisation of Australian Tribes*, 1931

Salter, E., *Daisy Bates*, 1983

Chapter 5: Malinowski

Darnell, R., *Readings in the History of Anthropology*, New York, 1974

Firth, R., ed., *Man and Culture: An Evaluation of the Work of Malinowski*, New York, 1957, especially Firth's 'Malinowski as Scientist and Man' and Leach's 'The Epistemological Background to Malinowski's Empiricism'

Geertz, C., 'Under the Mosquito Net', *New York Review of Books*, 14 September 1967

Kuper, A., *Anthropologists and Anthropology: The British School, 1922–1972*, 1996 (1973)

Malinowski, B., *A Diary in the Strict Sense of the Term*, 1967

——*Argonauts of the Western Pacific: An Account of Native Enterprise and Adventure in the Archipelagos of Melanesian New Guinea*, 1922

——*The Sexual Life of Savages*, 1968 (1929)

——*Crime and Custom in Savage Society*, 1926

——*Coral Gardens and their Magic*, 1935

Mauss, M., *The Gift*, trans. and annotated by J. I. Guyer, Chicago, 2016

Powdermaker, H., 'Further Reflections on Lesu and Malinowski's Diary', *Oceania* 40 (4), 1970

Spiro, M. E., *Oedipus in the Trobriands*, Chicago, 1982

Tilley, H., and R. J. Gordon, eds, *Ordering Africa: Anthropology, European Imperialism, and the Politics of Knowledge*, Manchester, 2007

Wayne, H. M., *The Story of a Marriage: The Letters of Bronislaw Malinowski and Elsie Masson*, 2 vols, 1995 and 2002

Wax, M., 'Tenting with Malinowski', *American Sociological Review* 37, 1972

Young, M., *Malinowski: Odyssey of an Anthropologist 1884–1920*, New Haven, 2004

——*Malinowski's Kiriwina: Fieldwork Photography 1915–1918*, Chicago, 1998

Young, M., ed., *Malinowski Among the Magi: The Natives of Mailu*, 1988

Chapter 7: Benedict

Babcock, B. A., and N. J. Parezo, *Daughters of the Desert: Women Anthropologists and the American Southwest, 1880–1980*, Albuquerque, 1988

Banner, L. W., *Intertwined Lives: Margaret Mead, Ruth Benedict and their Circle*, New York, 2003

Barnow, V., 'Ruth Benedict: Apollonian and Dionysian', *University of Toronto Quarterly* 18 (3), 1949

Belo, J., *Trance in Bali*, New York, 1960

Benedict, R., *Patterns of Culture*, New York, 1934

——*Zuni Mythology*, 1935

——*The Chrysanthemum and the Sword*, 1946

——*Race and Racism*, 1942

——'The Science of Custom', *Century*, April 1929

——'Anthropology and the Abnormal', *Journal of General Psychology* 10 (1), 1932

Deacon, D., *Elsie Clews Parsons: Inventing Modern Life*, 1997

Helm, J., *Pioneers of American Anthropology*, 1966

Jung, C. J., *Memories, Dreams, Reflections*, 1983 (1961)

Lavender, C. J., *Scientists & Storytellers: Feminist Anthropologists and the Construction of the American Southwest*, Albuquerque, 2006

Mead, M., ed., *Ruth Benedict: An Anthropologist at Work*, 2011 (1959)

——*Ruth Benedict: A Humanist in Anthropology*, New York, 2005 (1974)

Parezo, N. J., ed., *Hidden Scholars: Women Anthropologists and the Native American Southwest*, Albuquerque, 1993

Parsons, E. C., *Fear and Conventionality*, 1914

Parsons, E. C., ed., *American Indian Life*, New York, 1922

Sapir, E., 'Observations on the Sex Problem in America', *American Journal of Psychiatry* 8, 1928

Schachter, J., *Ruth Benedict: Patterns of a Life*, Philadelphia, 1983

Also see Benedict's chapter on 'Religion' in Boas, *General Anthropology*.

Chapter 8: Mead

Bateson, M. C., *With a Daughter's Eye: A Memoir of Margaret Mead and Gregory Bateson*, New York, 1984

Caffrey, M. M., and P. S. Francis, eds, *To Cherish the Life of the World: Selected Letters of Margaret Mead*, Cambridge, MA, 2006

Freeman, D., *The Fateful Hoaxing of Margaret Mead: A Historical Analysis* of her Samoan Research, Boulder, 1999

Lutkehaus, N. C., *Margaret Mead: The Making of an American Icon*, Princeton, 2008

Mead, M., *From the South Seas: Studies of Adolescence and Sex in Primitive Societies*, New York, 1939

——*Blackberry Winter: My Earlier Years*, 1973

——*Letters from the Field*, 1977

——*Sex and Temperament in Three Primitive Societies*, 1950

——*Male & Female: A Study of the Sexes in a Changing World*, 1949

Mead, M., and R. Métraux, *A Way of Seeing*, New York, 1970 (the collected *Redbook* articles)

Navoll, R., and R. Cohen, *A Handbook of Method in Cultural Anthropology*, 1970

An internet search suggests that Brand's anecdote is unsubstantiated, although it sounds highly likely.

Chapter 9: Hurston

Baker, L. D., *From Savage to Negro: Anthropology and the Construction of Race, 1896–1954*, Berkeley, 1998

Bloom, H., ed., *Zora Neale Hurston*, New York, 1986

Boyd, V., *Wrapped in Rainbows: The Life of Zora Neale Hurston*, 2003

Hurston, Z. N., *Dust Tracks on a Road*, 1944

——*Mules and Men*, 1936

——*Tell My Horse: Voodoo and Life in Haiti and Jamaica*, 1990

——*Barracoon: The Story of the last 'Black Cargo'*, ed. D. G. Plant, New York, 2018

——*Their Eyes Were Watching God*, 1937

Kaplan, C., *Zora Neale Hurston: A Life in Letters*, New York, 2002

Chapter 10: Richards

Asad, T., ed., *Anthropology & the Colonial Encounter*, 1973 (including essay by Wendy James)

Bohannan, L., 'Shakespeare in the Bush' (available online)

Bowen, E. S., *Return to Laughter*, 1954

Firth, R., 'Audrey Richards 1899–1984', *Man* 20 (2), June 1985

Gladstone, J., 'Significant Sister: Autonomy and Obligation in Audrey Richards' Early Fieldwork', *American Ethnologist* 13 (2), 1986

La Fontaine, J. S., ed., *The Interpretation of Ritual: Essays in Honour of A. I. Richards*, 1972

——'Audrey Isobel Richards, 1899–1984: An Appreciation', *Africa* 55 (2), 1985

Lamb, C., *The Africa House: The True Story of an English Gentleman and his African Dream*, 2000 (1999)

Lutkehaus, N., 'She was *very* Cambridge', *American Ethnologist* 13 (4), 1986

Richards, A., *Hunger and Work in a Savage Tribe*, 1932

——*Land, Labour and Diet in Northern Rhodesia*, 1939

———'Anthropological Problems in North-Eastern Rhodesia', *Africa* 5 (2), 1932

———'The Village Census in the Study of Culture Contact', *Africa* 8 (1), 1935

Shils, E., and C. Blacker, eds, *Cambridge Women: Twelve Portraits*, Cambridge, 1996

Issue 10 (1) of the *Cambridge Journal of Anthropology* (1985) was dedicated to Richards' memory, including pieces by Jean La Fontaine, Raymond and Rosemary Firth and Helena Malinowska Wayne. A recording of Richards in conversation with Dr Jack Goody in 1982 is available on YouTube. Her photos are in the Royal Anthropological Institute collection.

Frances Larson's poignant *Undreamed Shores: The Hidden Heroines of British Anthropology* (2021) vividly explores the lives of five female fieldworkers based in Oxford at the start of the twentieth century. In some ways it is the reverse of this book: where I have outlined the lives of well-known fieldworkers, Larson has rescued her subjects from obscurity.

Chapter 11: Lévi-Strauss

Cressy-Marcks, V., *Up the Amazon and Over the Andes*, 1932

Diamond, S., ed., *Culture in History: Essays in Honour of Paul Radin*, New York, 1960

———*In Search of the Primitive*, New Brunswick, 1974

Hayes, E. N. and T., eds, *Claude Lévi-Strauss: The Anthropologist as Hero*, Cambridge, MA, 1970

Lévi-Strauss, C., trans. J. and D. Weightman, *Tristes Tropiques*, Harmondsworth, 1984 (first pub. 1955 in French, this trans. 1973)

———*Wild Thought*, trans. J. Mehlman and J. Leavitt, Chicago, 2021

Sontag, S., review of *Structural Anthropology* in *The New York Review of Books*, 28 November 1963

Wilcken, P., *Claude Lévi-Strauss: The Poet in the Laboratory*, 2010

The book I found most helpful for deciphering Lévi-Strauss's academic work, which Edmund Leach comfortably describes as combining 'baffling complexity with overwhelming erudition', is his slim volume of 1970, *Lévi-Strauss*.

Conclusion

Comaroff, J. and J., *Ethnography and the Historical Imagination*, Boulder, 1992

Geertz, C., *The Interpretation of Cultures: Selected Essays*, 1973

Works and Lives: The Anthropologist as Author, Cambridge, 1988

Rubinstein, R. A., ed., *Doing Fieldwork: The Correspondence of Robert Redfield and Sol Tax*, New Brunswick, 2001

More General Reading

Augé, M., *A Sense for the Other: The Timeliness and Relevance of Anthropology*, trans. A. Jacobs, Stanford, 1998 (1994)

Barley, N., *The Innocent Anthropologist: Notes from a Mud Hut*, Harmondsworth, 1983

Boorstein, D., *The Discoverers*, 1983

Bunzel, R., and M. Mead, *The Golden Age of American Anthropology*, New York, 1960

Curtin, P., *The Image of Africa: British Ideas and Actions, 1780–1850*, Madison, 1964

D'Souza, D., *The End of Racism: Principles for a Multi-Racial Society*, New York, 1995

Evans-Pritchard, E. E., *Essays in Social Anthropology*, 1962

Fagan, B., ed., *The Great Archaeologists*, 2014

Gambrel, A., *Women Intellectuals, Modernism and Difference: Transatlantic Culture*, Cambridge, 1997

Gilkeson, J. S., *Anthropologists and the Rediscovery of America, 1886–1965*, Cambridge, 2010

Hobsbawm, E., *The Age of Extremes: A History of the World, 1914–1991*, 1996

Leeds-Hurwitz, W., *Rolling in Ditches with Shamans*, Lincoln, 2004

Lowie, R., *Primitive Society*, 1919
——*Are We Civilised?*, 1929

Lynd, R. and H., *Middletown*, 1929
——*Middletown in Transition: A Study in Cultural Conflicts*, 1937

Montagu, A., *The Practice of Love*, Englewood Cliffs, 1975

Myrdal, G., *An American Dilemma: The Negro Problem and Modern Democracy*, 1944

Powdermaker, H., *Hollywood, the Dream Factory: An Anthropologist Looks at the Movie Makers*, Boston, 1950

Rippon, G., *The Gendered Brain: The New Neuroscience that Shatters the Myth of the Female Brain*, 2019

Said, E., *Orientalism*, Harmondsworth, 1978

Stocking, G. W., *Malinowski, Rivers, Benedict and others: Essays on Culture and Personality*, Madison, 1986
——*Observers Observed: Essays on Ethnographic Fieldwork*, Madison, 1983
——*The Ethnographer's Magic and Other Essays*, Madison, 1992
——'The History of Anthropology', *Journal of the History of Behavioural Sciences* 2, 1966

Tilley, C., ed., *Reading Material Culture*, Oxford, 1990

Illustrations

Illustrations

Every effort has been made to trace or contact all copyright-holders. The publishers will be pleased to make good any omissions or rectify any mistakes brought to their attention at the earliest opportunity.

Index

Index

Index

Index

Index

Index

Index

Index

Index

Rio Tibagy, 253–5
Rio Xingu, 260n
Rishbeth, Kathleen, 50
Rising Tide of Color Against White World Supremacy, The (Stoddard), 146
Ritual and Belief in Morocco (Westermarck), 77
Rivers, William, 42, 46–65, 69, 85, 208, 236, 240
anthropometric measurements, views on, 213–14
Australian Aboriginals, study of, 91
British Association meeting (1914), 117
Carnegie Institute report (1913), 58–60
colonialism, views on, 58, 59, 85
colour perception studies, 49
death (1922), 63–4
diffusionism, 57–8, 64, 65, 147
double-blind procedure, 56
genealogical method, 49
History of Melanesian Society (1914), 56–8, 60
Instinct and the Unconscious (1920), 61
kinship theory, views on, 50, 54, 57, 65
Labour candidacy (1922), 63
nerves, study of, 56
racial suicide, views on, 58, 62
Radcliffe-Brown, relationship with, 93, 95
shell shock, study of, 60, 61
sleep deprivation, study of, 56
Solomon Islands expedition (1908), 56
string figures, views on, 50
Toda expedition (1901–2), 52–5, 56, 59, 64, 86, 141

Torres Strait expedition (1898), 42, 46–51, 116
Trobriand Islands expedition (1908), 119
unconscious, study of, 61
World War I (1914–18), 60, 62
Rivet, Paul, 258, 270, 272
Roman, Monique, 273
Rondon, Cândido Mariano da Silva, 259–60, 263
Roosevelt, Eleanor, 279
Rosetta Stone, 9–10
Rousseau, Jean-Jacques, 7, 9
Royal Anthropological Institute, 251
Royal Anthropological Society, 65
Royal Anthropological Society of Australasia, 101
Royal Ethnological Museum, Berlin, 13, 33
Royal Flying Corps, 60
Royal Geographical Society, 71, 261
Royal Geographical Society of Australasia, 100
Royal Society of Arts, 246
Royal Society of South Australia, 119
Rubinstein, Robert, 281
rum, 220, 277
Rupert Street market, London, 249
Russell, Bertrand, 63
Russell, Frank, 147
Russia, 7, 77

Samoa, 17, 191, 195–204
sexuality in, 198–204, 208–9
taupou, 201
San Ildefonso, New Mexico, 169, 175
Sandstone, Western Australia, 105
Sanger, Margaret, 138, 171, 202

Santa Cruz, Hernán, 279
Santo Domingo, Dominican Republic,174, 276
São Jeronymo, Brazil, 254
Sapir, Edward, 154, 155, 156, 187–8, 193, 194–5, 196, 203
Sassoon, Siegfried, 60, 62, 64
Saturday Review, 192
de Saussure, Ferdinand, 271
'savages', *see* 'primitiveness'
scalping, 187
Schmerler, Henrietta, 168n
Schumann, Robert, 22
Science, 34
scientific racism, 12–13, 34–5, 146–50, 158, 162
Scottish Nationalist Party, 71
self-mutilation, 178
Seligman, Charles, 46, 56, 116, 117, 119, 120, 130, 241
Seraph on the Suwanee (Hurston), 234
Sex and Repression in Savage Society (Malinowski), 203
Sex and Temperament in Three Primitive Societies (Mead), 205–8
Sexual Life of Savages, The (Malinowski), 130, 135–8
sexuality, 160, 180
homosexuality, 82–3, 137, 160, 173, 183, 208
Native Americans, 178, 179–82
New Guineans, 123–4, 125, 135–9, 205–8
promiscuity, 83, 136, 160, 199, 200
Samoans, 198–204, 208–9
United States, 202–3
Seychelles, 248
shamanism, 50
Shand, Alexander, 85
Shankland, David, 133
Sheen, Herbert, 220–21
Sheldrake, Merlin, 283
shell shock, 60, 61

Index

Index